GREEN ENERGY

GREEN ENERGY
A Sustainable Future

M. A. PARVEZ MAHMUD
School of Electrical, Mechanical and Infrastructure Engineering
The University of Melbourne
Parkville, VIC, Australia

SHAHJADI HISAN FARJANA
School of Engineering
Deakin University
Geelong, VIC, Australia

CANDACE LANG
School of Engineering
Macquarie University
Sydney, NSW, Australia

NAZMUL HUDA
School of Engineering
Macquarie University
Sydney, NSW, Australia

Academic Press is an imprint of Elsevier
125 London Wall, London EC2Y 5AS, United Kingdom
525 B Street, Suite 1650, San Diego, CA 92101, United States
50 Hampshire Street, 5th Floor, Cambridge, MA 02139, United States
The Boulevard, Langford Lane, Kidlington, Oxford OX5 1GB, United Kingdom

Copyright © 2023 Elsevier Inc. All rights reserved.

MATLAB® is a trademark of The MathWorks, Inc. and is used with permission.
The MathWorks does not warrant the accuracy of the text or exercises in this book.

This book's use or discussion of MATLAB® software or related products does not constitute endorsement or sponsorship by The MathWorks of a particular pedagogical approach or particular use of the MATLAB® software.

No part of this publication may be reproduced or transmitted in any form or by any means, electronic or mechanical, including photocopying, recording, or any information storage and retrieval system, without permission in writing from the publisher. Details on how to seek permission, further information about the Publisher's permissions policies and our arrangements with organizations such as the Copyright Clearance Center and the Copyright Licensing Agency, can be found at our website: www.elsevier.com/permissions.

This book and the individual contributions contained in it are protected under copyright by the Publisher (other than as may be noted herein).

Notices

Knowledge and best practice in this field are constantly changing. As new research and experience broaden our understanding, changes in research methods, professional practices, or medical treatment may become necessary.

Practitioners and researchers must always rely on their own experience and knowledge in evaluating and using any information, methods, compounds, or experiments described herein. In using such information or methods they should be mindful of their own safety and the safety of others, including parties for whom they have a professional responsibility.

To the fullest extent of the law, neither the Publisher nor the authors, contributors, or editors, assume any liability for any injury and/or damage to persons or property as a matter of products liability, negligence or otherwise, or from any use or operation of any methods, products, instructions, or ideas contained in the material herein.

ISBN: 978-0-323-85953-0

For information on all Academic Press publications
visit our website at https://www.elsevier.com/books-and-journals

Publisher: Joseph P. Hayton
Acquisitions Editor: Lisa Reading
Editorial Project Manager: Moises Carlo P. Catain
Production Project Manager: Kamesh Ramajogi
Cover Designer: Miles Hitchen

Typeset by VTeX

Contents

List of figures ix
List of tables xiii

1. **Introduction to green and sustainable energy** 1
   1.1. Challenges and objectives 1
   1.2. Main contributions 3
   1.3. Book outline 5

2. **State-of-the-art life cycle assessment methodologies applied in renewable energy systems** 7
   2.1. Introduction 7
   2.2. Review selection criteria and method 11
   2.3. Life cycle assessment of renewable power plants 12
   2.4. LCA of renewable energy systems 17
   2.5. Geographic location-wise LCA of renewable energy systems 31
   2.6. Summary and outlook 40
   2.7. Conclusion and future recommendation 45

3. **Environmental impacts of solar-PV and solar-thermal plants** 47
   3.1. Introduction 48
   3.2. Materials and methods 52
   3.3. Results and discussion 58
   3.4. Limitations of this study 71
   3.5. Conclusions 71

4. **Environmental impacts of hydropower plants** 73
   4.1. Introduction 73
   4.2. Hydropower plants of alpine and nonalpine areas in Europe 78
   4.3. Methodology 80
   4.4. Results 88
   4.5. Discussion 96
   4.6. Limitations and future improvements 98
   4.7. Conclusion 102

v

5. **Environmental impact assessment of renewable power plants in the US** — 103
   - 5.1. Introduction — 103
   - 5.2. US electricity generation and consumption overview — 109
   - 5.3. Methodology — 109
   - 5.4. Results and interpretation — 118
   - 5.5. Uncertainty analysis — 124
   - 5.6. Sensitivity analysis — 129
   - 5.7. Discussion — 131
   - 5.8. Conclusion — 132

6. **Comparative environmental impact assessment of solar-PV, wind, biomass, and hydropower plants** — 135
   - 6.1. Introduction — 136
   - 6.2. Materials and methods — 138
   - 6.3. Results and discussion — 147
   - 6.4. Conclusion — 160

7. **Advanced energy-sharing framework for robust control and optimal economic operation of an islanded microgrid system** — 161
   - 7.1. Introduction — 161
   - 7.2. Power-routing framework — 164
   - 7.3. Optimization-based energy-sharing model — 166
   - 7.4. Power-routing control strategy — 168
   - 7.5. Simulation and results — 171
   - 7.6. Conclusion — 177

8. **Environmental impact assessment and techno-economic analysis of a hybrid microgrid system** — 179
   - 8.1. Introduction — 179
   - 8.2. Microgrid system overview — 183
   - 8.3. Methods — 185
   - 8.4. Results and discussion — 193
   - 8.5. Sensitivity analysis — 198
   - 8.6. Conclusion — 200

9. **Future directions towards green and sustainable energy** — 205
   - 9.1. Book summary and concluding remarks — 205
   - 9.2. Future research directions — 209

A. List of acronyms	**211**
B. List of symbols	**213**
References	*215*
Index	*231*

List of figures

Figure 2.1	Systematic overview of the key steps followed to conduct this review.	12
Figure 2.2	The key LCA stages [1].	12
Figure 2.3	The LCA framework [1].	13
Figure 2.4	The common life cycle inventory for energy systems [2].	14
Figure 2.5	Schematic representation of LCA methods.	15
Figure 2.6	Comparison of key impacts of various renewable plants [2].	43
Figure 2.7	Key impacts of solar-PV plants.	43
Figure 2.8	Key impacts of hydropower plants.	44
Figure 2.9	Key impact comparison with wind power plants.	44
Figure 2.10	Key impact comparison with biomass power plants.	45
Figure 3.1	Schematic framework of the solar-PV system.	53
Figure 3.2	Schematic framework of the solar-thermal system.	53
Figure 3.3	Step-by-step energy and material flows for both systems.	55
Figure 3.4	System boundary of the LCA.	56
Figure 3.5	Life cycle inputs and outputs of the solar-PV system using the RMF methodology.	59
Figure 3.6	Environmental profiles of the considered solar-PV system.	60
Figure 3.7	End-point impacts of the individual components of the solar-PV system.	60
Figure 3.8	Life cycle inputs and outputs of the solar-thermal system using the RMF methodology.	62
Figure 3.9	Environmental profiles of the considered solar-thermal system.	62
Figure 3.10	End-point impacts of the individual components of the solar-thermal system.	63
Figure 3.11	Comparison of environmental impacts from the solar-PV and the solar-thermal system.	64
Figure 3.12	End-point impact comparison of the systems using Impact 2002+ methodology.	65
Figure 3.13	GHG emission of the solar-PV system with a time period of 100 years.	66
Figure 3.14	GHG emission of the solar-thermal system with a time period of 100 years.	67
Figure 3.15	GHG emission of the systems as determined using IPCC methodology.	67
Figure 3.16	Required energy from different sources to build, operate, and dispose of both systems.	68

Figure 3.17	Probability distribution for the single-score impact category of the solar-PV system.	70
Figure 3.18	Probability distribution for the single-score impact category of the solar-thermal system.	71
Figure 4.1	Map of the alpine boundary in Europe (source: 2nd Report on the State of the Alps) [3].	79
Figure 4.2	Hydropower production scenarios in alpine and nonalpine areas of Europe [4].	80
Figure 4.3	Stages of the LCA method [5].	81
Figure 4.4	Materials flow sheet for 1 MJ of hydropower generation in an alpine region.	82
Figure 4.5	Materials flow sheet for 1 MJ of hydropower generation in a nonalpine region.	83
Figure 4.6	LCA system boundary used in this research.	84
Figure 4.7	LCA methods used in this research.	87
Figure 4.8	Global warming-based impact outcome comparison.	89
Figure 4.9	Ozone formation-based impact outcome comparison.	89
Figure 4.10	Ecotoxicity-based impact outcome comparison.	90
Figure 4.11	Water consumption-based impact outcome comparison.	90
Figure 4.12	Effect outcome comparison for other impact indicators.	91
Figure 4.13	End-point damage assessment of the plants using the Impact 2002+ approach.	92
Figure 4.14	GHG emissions as determined by the IPCC approach.	94
Figure 4.15	Comparative life cycle inputs and outputs of hydropower plants of alpine and nonalpine regions as determined by the RMF method.	95
Figure 4.16	Environmental impacts of various power plants.	99
Figure 4.17	Probability distribution for the single-score impact category of hydropower plants of alpine zones.	99
Figure 4.18	Probability distribution for the single-score impact category of hydropower plants of nonalpine zones.	99
Figure 5.1	Electricity consumption overview in the US based on different energy sources [6].	111
Figure 5.2	Material flow sheet for 1 kWh of solar-PV power generation.	112
Figure 5.3	Material flow sheet for 1 kWh of pumped storage hydropower generation.	113
Figure 5.4	Material flow sheet for 1 kWh of biomass power generation.	114
Figure 5.5	Common system boundary for all power generation processes used in this LCA analysis.	116
Figure 5.6	Life cycle impact assessment methods used in this research.	117

List of figures

Figure 5.7	Normalized environmental impact outcomes, as determined using the TRACI mid-point approach.	119
Figure 5.8	LCA outcome after weighting by the Eco-indicator 99 end-point approach.	121
Figure 5.9	Metal- and gas-based emissions by renewable energy plants as determined using the Eco-points 97 method.	122
Figure 5.10	GHG emissions as determined using IPCC methodology.	124
Figure 5.11	Probability distribution for the single-score impact category of the solar-PV power plant.	127
Figure 5.12	Probability distribution for the single-score impact category of the pumped storage hydropower plant.	128
Figure 5.13	Probability distribution for the single-score impact category of the biomass power plant.	128
Figure 5.14	Comparison of the findings with existing studies.	132
Figure 6.1	Material flow sheet for 1 MJ of solar energy generation.	141
Figure 6.2	Material flow sheet for 1 MJ of wind energy generation.	142
Figure 6.3	Material flow sheet for 1 MJ of hydro energy generation.	143
Figure 6.4	Material flow sheet for 1 MJ of biomass energy generation.	144
Figure 6.5	Common system boundary for all power generation processes used in this LCA analysis.	145
Figure 6.6	Life cycle impact assessment methods used in this research.	146
Figure 6.7	Comparative LCA inputs and outputs of the considered renewable energy plants using the RMF method.	148
Figure 6.8	Comparison per impact indicator using CML mid-point methodology by adding individual effects. The highest impact is set to 100%.	149
Figure 6.9	LCA outcomes after weighting by the Eco-indicator 99 end-point approach.	150
Figure 6.10	Relative fuel-based energy consumption rates by the considered plants, as determined using the CED method.	151
Figure 6.11	Relative GHG emissions by the plants, as determined using the IPCC methodology.	153
Figure 6.12	Probability distribution for the single-score impact category of the PV power plant.	154
Figure 6.13	Probability distribution for the single-score impact category of the wind power plant.	154
Figure 6.14	Probability distribution for the single-score impact category of the hydropower plant.	155
Figure 6.15	Probability distribution for the single-score impact category of the biomass power plant.	155
Figure 6.16	Outcome comparisons with prior studies.	156

Figure 7.1	The conceptual architecture for the power-routing framework.	166
Figure 7.2	Power routing management strategy.	167
Figure 7.3	Inverter control structure.	170
Figure 7.4	Inverter circuit diagram with LCL filter.	170
Figure 7.5	Aggregated PV generation and load profile of prosumers and consumers.	172
Figure 7.6	Individual load profiles of prosumers.	172
Figure 7.7	Individual load profiles of consumers.	173
Figure 7.8	Solar irradiation profile at the MG location.	173
Figure 7.9	Hourly prosumers' demand and supply status.	174
Figure 7.10	Hourly consumers' demand and supply status.	175
Figure 7.11	%SoC of the CSS.	175
Figure 7.12	Hourly profit from the MG framework.	176
Figure 7.13	AC and DC bus voltages.	177
Figure 7.14	AC bus frequency.	177
Figure 8.1	The MG framework structure.	184
Figure 8.2	The system boundary of the MG framework for LCA analysis.	189
Figure 8.3	The stage-wise material, energy, and emission flow.	190
Figure 8.4	The material flow of the MG framework.	191
Figure 8.5	The LCA methods used in this analysis.	192
Figure 8.6	The annual excess power rate of the MG framework.	194
Figure 8.7	The life cycle environmental profiles of the framework as determined using the ReCiPe 2016 method.	195
Figure 8.8	End-point damage assessment of the framework using the ReCiPe 2016 method.	196
Figure 8.9	GHG emission as determined using the IPCC method.	198
Figure 8.10	The metal-based emissions quantification outcome using the Eco-points 97 method.	199

List of tables

Table 2.1	The best practice method for each impact indicator [7].	16
Table 2.2	Key findings and recommendations from recent studies on LCA of solar-PV plants.	19
Table 2.3	Key findings and recommendations from recent studies on LCA of hydropower plants.	25
Table 2.4	Key findings and recommendations from recent studies on LCA of wind power plants.	28
Table 2.5	Key findings and recommendations from recent studies on LCA of biomass power plants.	30
Table 2.6	Key findings and recommendations from recent studies on LCA of other renewable plants.	32
Table 2.7	Key findings and recommendations from recent studies on LCA of renewable plants in Asia.	34
Table 2.8	Key findings and recommendations from recent studies on LCA of renewable plants in Europe.	36
Table 2.9	Key findings and recommendations from recent studies on LCA of renewable plants in America.	38
Table 2.10	Key findings and recommendations from recent studies on LCA of renewable plants in other zones.	39
Table 2.11	Comparison of key mid-point impacts among various renewable plants.	42
Table 3.1	Previous works of life cycle assessment of solar-thermal systems and their limitations.	50
Table 3.2	Previous works of life cycle assessment of solar-PV system and their limitations.	51
Table 3.3	Data collection for frameworks in the solar-PV system and the solar-thermal system.	57
Table 3.4	Life cycle inputs and outputs comparison between the solar-PV system and the solar-thermal system.	64
Table 3.5	Sensitivity analysis outcome for different solar collector types for the solar-thermal system.	69
Table 3.6	Sensitivity analysis outcome based on different battery types for the solar-PV system.	70
Table 4.1	Recent studies on LCA of hydropower plants and the research gaps.	76
Table 4.2	Hydropower production details for the alpine areas in Europe.	79
Table 4.3	Hydropower production details for the nonalpine areas in Europe.	79

Table 4.4	LCI for LCA of the considered hydropower plants located in alpine regions.	85
Table 4.5	LCI for LCA of the considered hydropower plants located in nonalpine regions.	86
Table 4.6	Life cycle energy consumption by the considered hydropower plants, as determined by the CED method.	95
Table 4.7	Key impact comparison with previous studies.	96
Table 4.8	Key impacts of various plants.	100
Table 4.9	Key damage comparison with various plants.	101
Table 5.1	Country-based overview of previous research on solar-PV, hydro, and biomass power plants.	106
Table 5.2	Electricity production in the US based on different energy sources [8].	110
Table 5.3	Data sources for the considered renewable power plants in the US.	115
Table 5.4	LCA inputs and outputs of the considered plants using the RMF approach.	119
Table 5.5	Mid-point environmental impacts of the considered plants as determined using the TRACI method.	120
Table 5.6	Fuel-based energy consumption rates of solar-PV, pumped storage hydropower, and biomass plants in the US.	121
Table 5.7	Plants' metal- and gas-based emissions as determined by the Eco-points 97 method.	123
Table 5.8	GHG emissions as determined by the IPCC approach.	123
Table 5.9	Mid-point impact comparison with other nonrenewable power plants.	125
Table 5.10	End-point impact comparison with other nonrenewable power plants.	126
Table 5.11	Sensitivity analysis of seven pumped storage hydropower plants in various countries.	130
Table 6.1	Data sources for the considered renewable power plants.	140
Table 6.2	Metal- and gas-based emissions by renewable energy plants, as determined using the Eco-points 97 method.	152
Table 6.3	Important findings and comparison with other studies.	157
Table 7.1	Control system parameters.	171
Table 8.1	Simulation parameters.	187
Table 8.2	Data collection for LCA of the MG framework.	190
Table 8.3	The NPC-based optimization result of the MG framework.	194
Table 8.4	The key hazardous substances of the MG elements that mostly affect the end-point environmental indicators.	197

Table 8.5	Sensitivity analysis outcomes for various battery lifetimes and solar-scaled factors in NPC-based optimization of the MG.	200
Table 8.6	Sensitivity analysis outcome for various PV modules of the MG.	201
Table 8.7	Sensitivity analysis outcome for various batteries of the MG.	202

CHAPTER ONE

Introduction to green and sustainable energy

The demand for electricity is increasing day by day due to the rise of the global population and the increasing industrial activity. To fulfill this increasing energy demand, more carbon-based fuels are being used in conventional power plants, which release greater amounts of greenhouse gases (GHGs) and hazardous substances into the environment. The generation of renewable electricity, however, for example with solar-photovoltaic (PV), wind, biomass, and hydropower plants, can replace the fossil fuel-based production in a sustainable way as they produce low-carbon electricity through techno-economic operation. Thus, it helps to fulfill the energy demands via an environment-friendly and economically viable approach. However, like conventional energy systems, renewable power plants also have some direct and indirect impacts on the environment, human health, ecosystems, and resources, which come from each element's production, transportation, installation, operation, and end-of-life recycling stages. Therefore, it is needed to assess the environmental impact and economic viability of renewable energy plants and compare them with those of other options for a region based on a dynamic life cycle assessment (LCA) approach. In this book we aim to identify the processes of renewable energy systems that contribute most to environmental and economic benefits.

1.1. Challenges and objectives

The main challenges of this research work are listed as follows:
- The life cycle environmental impacts of all elements of renewable plants like solar-PV, solar-thermal, and hydropower generation systems must be identified by an appropriate LCA approach to replace the impacting materials by alternative environment-friendly options.
- The development of comprehensive life cycle inventory (LCI) approaches considering all stages of renewable energy plants from raw material extraction to end-of-life waste management is necessary for assessing the environmental impacts.

- The quantification of GHG emissions by renewable plants located in different geographic locations at each stage of their life cycle is essential to find cleaner options.
- The estimation of fossil fuel-based energy consumption over the lifetime of renewable power plants including the raw material extraction, manufacture of key parts, transportation, system installation, and end-of-life waste disposal stages is needed to replace carbon-based systems by sustainable approaches.
- It is essential to evaluate the metal-based emissions into air, water, and soil over the lifetime of renewable plants by an appropriate LCA method to save the environment and achieve cleaner electricity production.
- Uncertainty and sensitivity analyses of renewable power plant elements are required to optimize their operation with respect to environmental impact and economic gain. It is also required to obtain information on the environmental impact of each element of the considered renewable energy systems and on ways to supply electricity with cleaner generation, higher cost-effectiveness, and higher reliability.
- The development of a renewable energy-driven microgrid (MG) to share excess electricity in a community for the optimal use of distributed energy resources and battery energy storage systems through nonlinear programming (NLP)-based optimization approaches is essential. It is also necessary to verify the proposed framework by a droop controller-based real-time control strategy for stable direct current (DC) and alternating current (AC) bus voltages to ensure better performance of the grid.
- It is required to conduct net present cost (NPC)-based energy optimization for optimal sizing of a solar PV-driven islanded MG using real-time physical, operation, and economic inputs in the system. It is also required to develop an LCI for this MG framework to evaluate the environmental impacts based on 21 mid-point indicators, 3 endpoint indicators, GHG emissions, and fossil fuel-based energy consumption over its lifetime using different systematic LCA approaches.

Motivated by the challenges related to the sustainable, clean, and economic operation of renewable energy technologies, the main objectives of the book are as follows:
- developing a comprehensive system boundary for solar-PV and solar-thermal systems for assessment, comparison, and sensitivity analysis of

their life cycle impacts on the environment, human health, the climate, and ecosystems;
- developing a unique LCI and quantifying the environmental impacts of hydropower plants in alpine and nonalpine areas of Europe through LCA analysis for identifying the best option;
- quantifying the environmental hazards of renewable electricity generation systems in the US, as the US have notable solar-PV, biomass, and pumped storage hydropower plants, and comparing their effects on the environment and human health through a dynamic LCA approach;
- designing a novel LCI for solar-PV, wind, and hydropower plants in Switzerland to evaluate life cycle emissions and identify the best plant option;
- developing an advanced power-routing framework for a solar-PV-based islanded MG with a central storage system for optimal economic operation through routing excess energy to nearby neighborhoods;
- conducting an NPC-based simulation for optimal sizing of an islanded MG framework and assessing its environmental impacts based on 21 mid-point indicators and 3 end-point indicators by developing a novel LCI for the system.

1.2. Main contributions

Following the abovementioned research challenges and objectives of this book, the six key contributions are highlighted below:
- The first contribution of this book is the design, system development, data collection, and LCA-based evaluation of the environmental effects of a solar-PV and a solar-thermal system. To ensure effectiveness of this research, a comprehensive system boundary is developed for both considered solar technologies, LCA is carried out for both systems by multiple methods to assess the environmental profiles, the GHG emission rates are estimated for both systems, and sensitivity and uncertainty analysis is conducted to examine the environmental performance of both systems more critically.
- The second contribution of this book is the environmental hazard estimation of existing hydropower plants in Europe. For that reason, a unique LCI is developed for hydropower plants located in both alpine and nonalpine areas of Europe to assess their environmental hazards in terms of ecosystems, climate change, resources, and human health.

Moreover, a step-by-step LCA analysis is performed to determine the GHG and metal-based emissions of both categories of plants.
- The third contribution of this book is the comparative environmental impact assessment of three different renewable power plants, namely solar-PV, biomass, and pumped storage hydropower plants in the US. The impacts are considered based on 10 mid-point impact categories and 3 end-point indicators.
- The fourth contribution of this book is the design and development of a new LCI for solar-PV, wind, biomass, and hydropower plants to evaluate life cycle emissions and identify the best plant option.
- The fifth contribution of this book is the establishment an advanced power-routing framework for a solar-PV-based islanded MG. An NLP-based optimization model is developed and applied for the day-ahead scheduling of excess power routing for an optimal profit to stakeholders using the proposed MG framework. Moreover, a modified droop controller-based real-time control strategy is established that provides stable DC and AC bus voltages and ensures better performance of the grid. The proposed power-routing framework is verified via a case study for a typical islanded MG.
- The final contribution of this research is the accomplishment of an NPC-based optimization analysis and an LCA-based assessment of the environmental impact of an off-grid MG framework. To ensure the validity of this research, NPC-based optimization is carried out for the highest profit of prosumers through optimal sizing of elements using real-time physical, operation, and economic inputs in the proposed MG system. Moreover, a novel LCI is developed to evaluate the life cycle material and energy flow and compare the environmental impacts of each element of the MG.

The well-known SimaPro software and the renowned ecoinvent global database are used to assess the life cycle environmental impacts by multiple methods such as the International Life Cycle Data System (ILCD) for mid-point analysis, Impact 2002+ for end-point analysis, cumulative energy demand (CED) for fossil fuel-based energy consumption estimation, Eco-points 97 for metal- and gas-based emission assessment, Eco-indicator 99 for uncertainty analysis, and Intergovernmental Panel on Climate Change (IPCC) assessment for GHG emission evaluation. Additionally, the HOMER Pro and MATLAB® tools are used for NPC- and NLP-based optimization of the economic operation of the MG framework for the maximum profit of prosumers and robust control operation of the proposed system. The findings of this book provide valuable information on the environmental impacts of each element of the considered renewable energy systems and on ways to supply electricity with cleaner generation, higher cost-effectiveness, and higher reliability. Overall, the results reveal the production impacts and can be utilized to prioritize environment-friendly and cost-effective operation correctly through potential improvement plans.

1.3. Book outline

This book contains a general introduction, Plant-based and country-based LCA of renewable energy generation systems, environmental impact assessment, and economic analysis of islanded and hybrid energy MG frameworks. There are three major branches: LCA, economic analysis (EA), and combined LCA and EA. In the LCA branch, life cycle impact analyses of both plant-based (solar-PV, solar-thermal, and hydropower plants) and country-based (US) renewable energy plants are presented. Then, in the EA branch, NLP-based economic optimization of a proposed islanded MG framework is addressed. Finally, in the combined LCA and EA branch, the NPC-based economic optimization and life cycle impacts of a proposed hybrid MG framework are discussed.

This book is outlined as follows.

Chapter 1 provides the challenges and objectives of this research including the main contributions. The book outline is presented at the end of this chapter.

Chapter 2 focuses on the background information and related works of LCA and EA of renewable energy technologies. This chapter also discusses the methods of LCA and economic optimization.

Chapter 3 presents the LCA of a solar-PV and a solar-thermal power plant. This chapter also compares the findings of both plant categories to make better-informed choices. It provides additional information on the impacts associated with each of the elements like the PV panel, valve, battery, converter, controller, flow meter, etc., in both solar-PV and solar-thermal systems, identifying the critical materials and stages; it is anticipated that by replacing the hazardous materials the environment can be saved from long-term dangerous emissions.

Chapter 4 concentrates on the analyses of environmental impacts of hydropower plants in alpine and nonalpine areas of Europe by a systematic LCA approach. This chapter presents the role of hydropower in promoting sustainable production of electricity, especially using the full potentials of the alpine region, thus leading to environmentally friendly clean renewable energy generation.

Chapter 5 discusses the environmental effects caused by different types of renewable plants through LCA. A comparative study is conducted among solar-PV, biomass, and pumped storage hydropower plants in the US. Life cycle impact analysis has been carried out by the Eco-indicator 99, Tool for the Reduction and Assessment of Chemical and other Environmental Impacts, Raw Material Flows, CED, Eco-points 97, and IPCC methods, using SimaPro software. The impacts are considered based on 10 mid-point impact categories and 3 end-point indicators. The findings will guide investors in installing sustainable and clean power plants.

Chapter 6 highlights a comparative LCA analysis of solar-PV, wind, biomass, and hydropower plants, identifying the best renewable plant option considering environmental perspectives.

Chapter 7 proposes a novel energy-sharing framework for a remote locality where an MG is the only means of meeting the prosumers' load demands and routing excess energy to their neighbors to fulfill their energy demands. It also presents an NLP-based optimization model for the day-ahead scheduling of maximum available power routing after fulfilling prosumers' and consumers' load demands within the constraints of central storage for an off-grid remote MG framework. Furthermore, the proposed power-routing framework is verified in this chapter to be a stable control operation method with proper voltage regulation utilizing a droop controller in the power control loop of the inverter.

Chapter 8 provides an NPC-based optimization analysis and an LCA-based environmental impact assessment of a solar-PV-driven off-grid MG framework. This chapter highlights the research outcomes from six perspectives: (i) proposing an off-grid MG system, (ii) optimizing the system based on NPC minimization, (iii) analyzing life cycle material flow, (iv) building an LCI approach, (v) assessing environmental profiles by multiple methods, and (vi) conducting sensitivity analyses to optimize the design and environmental performance of the proposed system. The well-known HOMER Pro and SimaPro software programs and the renowned ecoinvent global database are used for the cost optimization and impact assessment.

Chapter 9 provides concluding remarks and proposes future directions for this research area.

CHAPTER TWO

State-of-the-art life cycle assessment methodologies applied in renewable energy systems

This chapter reviews the major research findings published on the environmental impact evaluation of renewable electricity generation systems through life cycle assessment (LCA). The chapter provides important insights into the knowledge gaps in low-carbon electricity production technologies for a sustainable future. The main research focus in most parts of the world is on renewable power generation plants because of their true potential for low-carbon, nonfossil energy production. However, these plants have some negligible environmental impacts on humankind, resources, and ecosystems. These impacts occur mostly during the extraction of raw materials, the production of elements from the extracted raw materials, the transportation of the materials to the plant, and end-of-life waste management. Among a few environmental impact estimation methods which are widely used to determine sustainability indicators, LCA is a well-justified method. Although state-of-the-art resources and tools have been employed recently in evaluating the environmental impacts of renewable power plants throughout their lifespan, there is still a research gap in identifying the key processes which require most attention. The review findings reveal that the assessment indicators of resources and ecosystems are key factors that are mostly lacking in the previous literature which are crucial for people or locations nearby plants. This chapter analyzes and summarizes the existing literature to identify the research gaps to guide future research in the field of sustainable renewable electricity generation.

2.1. Introduction

The demand for electricity is increasing day by day due to the rise of the global population and the increasing industrial activity. To meet the increasing demand, fossil fuel-based conventional electricity production has been augmented [9]. These conventional power plants emit increasing amounts of harmful greenhouse gases (GHGs) through the burning of carbon fuels such as coal, gas, and oil [10]. Therefore, the global climate is being affected. Researchers are getting more concerned about the resources and ecosystems, due to the inevitable threats of the release of harmful substances into the atmosphere. It is required to take steps to abate global GHG emissions to save the environment. Therefore, nowadays renewable energy systems (RESs) are getting more popular to abate the conventional carbon-based energy generation [11]. In the next decades, we will see an unprecedented expansion of RESs for electricity production. Solar-photovoltaic (PV), wind, biomass, and hydropower are the most promising RESs to be considered due to their high reliability and sustainability [12,13].

It is usually considered that renewable power technologies have smaller environmental impacts than conventional generation systems, but the impacts of RESs are not negligible. It is needed to quantify the environmental hazard of each element during the manufacture, transportation, installation, operation, maintenance, and end-of-life recycling processes of the solar-PV, wind, biomass, and hydropower plants. For that purpose, the LCA approach is widely used to evaluate the impacts of power plants throughout their lifespan in a step-by-step manner [14]. LCA follows the standard approach of the International Standardization Organization (ISO) [15] in assessing the environmental impacts with accuracy and robustness. LCA has been used in ample sustainability assessments of various products and systems. Lelek et al. reported that LCA is an important tool for environmental impact evaluation of energy systems as it considers life cycle input resources, material flows, and output emissions for overall impact quantifications [16]. Thus, comparative LCA results can be used as a theoretical basis to make better-informed choices and improve sustainability.

In the last 10 years, about 67 important research works have been published based on LCA of renewable energy technologies. The prior studies depict that the effects of a power plant vary based on its geographical location. For instance, the effects of hydropower plants located in the alpine regions of Europe differ from the effects of hydropower plants situated in nonalpine regions of Europe. Previous literature also suggested that the sustainability of renewable energy plants in a specific area depends on the abundance of resources in that zone. Moreover, the dominant elements that are responsible for most impacts of each power plants were highlighted in the previous studies. The main purpose of this critical review is to analyze and summarize recent findings from the LCA of RESs and provide a future direction to optimize the sustainability of renewable energy production.

A number of prior investigations highlighted the life cycle effects of one major element of the plant, like the PV panel, turbine, or battery. Some other studies focused on one type of plant in a specific location, like solar-PV, wind, or hydropower plant. There also exist prior works that assessed and compared various renewable plants of a country or specific location. Overall, researchers depicted mostly plant-based LCA and country-based LCA of RESs. Liang et al. [17], Akinyele et al. [18], and Mahmud et al. [19] conducted LCA of lithium-ion batteries. Innocenzi et al. [20] and Meng et al. [21] assessed the effects of NiMH batteries. Espinosa et al. [22,23], Latunussa et al. [24], and Gerbinet et al. [25] analyzed the impacts of solar-PV panels. Hou et al. [26], Atilgan et al. [27], Mahmud et al. [28], Ward et al. [29], Das et al. [30], Rubio et al. [31], Santoy et al. [32], and Jacobson et al. [33] evaluated the impacts of solar technologies. Srinivasan et al. [34], Gaudard et al. [35], Brioneshidrovo et al. [36], and Scherer et al. [37] evaluated the environmental impacts of hydropower technologies. Moreover, Turconi et al. [38], Huang et al. [39], Jesuina et al. [40], Xu et al. [41], Schreiber et al. [42], and Fang et al. [43] depicted the environmental effects of wind power technologies and identified sustainable ways of constructing wind turbines for superior environmental profiles. Furthermore, Beagle et al. [44], Pedro et al. [45], and Maier et al. [46] assessed the environmental impacts of biomass power plants and suggested the most influencing factors in bioenergy supply chains. Such plant-based LCA analyses of renewable energy technologies highlighted ways of lowering carbon emissions in the near future, which are summarized to present the state of the art and to provide future research directions. Some other previous studies only investigated the effects of plants in a specific location, like Asia [15,47–50], Europe [51–55], or America [31,36,56,57], and focused on finding the best renewable energy option in the zone considering sustainability. Such country-based LCA analysis of renewable energy technologies highlighted ways of sustainable renewable energy generation, but none analyzed the amount of fossil fuel-based energy consumption by the considered plants during construction, usage, and end-of-life management in different countries.

There are three review articles in this field. In 2019, Barros et al. presented a critical review after conducting a study considering 67 recent relevant articles. In this review, they analyze and summarize the characteristics of the literature, identifying the most used impact categories, LCA tools, keywords, journals, research groups, and their locations [14]. However, a summary of the technical outcomes and future directions has not been provided. In 2013, Turconi et al. reviewed LCAs of electricity generation technologies and demonstrated the environmental consequences of implementing new technologies based on varying existing LCA findings [38]. The major impacting elements of the plants have not been evaluated. In 2009, Varun et al. estimated and compared the carbon emissions of renewable energy generation systems [58], but the rate of fossil fuel-based energy consumption over the overall lifespan of the plants and ways to reduce carbon release into the environment has not been reported yet.

Though these environmental hazards differ widely between different plant types, between different locations, and between electricity generation systems of different sizes, it is necessary to estimate the impacts from each RES to recognize the main elements which are responsible for most environmental hazards/emissions. For instance, the dominant factors affecting the major impacts of a hydropower plant are construction, transport, the turbine, and the reservoir [11]. It is essential to analyze and compare the recent findings on GHG emissions by RESs to reduce global warming. The literature lacks a comprehensive analysis and comparison of fossil fuel-based energy consumption by RESs throughout their lifetime. Moreover, the impacts of the renewable systems on human health, ecosystems, the climate, and resources are not been summarized in prior works. Therefore, the main contribution of this chapter is sixfold. First, it highlights the LCA methods of RESs and indicates the issues that must be overcome. Second, it analyzes and summarizes the key findings and recommendations for the sustainable development of RESs in recent studies for further improvements. Third, it reviews the key impacting elements of each plant in different countries to replace them by an equivalent sustainable alternative. Fourth, it estimates the rate of fossil fuel-based energy consumption during the overall lifespan of the RES. Fifth, it compares the rate of GHG emission from various renewable power plants. Last, it analyzes and summarizes the impacts of RESs on human health, ecosystems, climate, and resources.

The structure of this chapter, including the main objectives, considerations, and contributions, are presented briefly in this **Section 2.1**. For the accomplishment of the key objectives of this study, **Section 2.2** describes the review selection criteria and approaches used in this work. **Section 2.3** highlights the LCA methodology used for impact assessment of renewable energy technologies. **Section 2.4** depicts the summary of sustainability outcomes of solar-PV, solar-thermal, wind, and hydropower plants. The findings of the life cycle impact assessment of RESs at different geographic locations are summarized in **Section 2.5**. **Section 2.6** summarizes and compares the key factors responsible for environmental hazards from RESs and indicates the research gaps in the prior LCA research in relation to renewable power systems. Lastly, concluding remarks and future research directions are presented in **Section 2.7**.

2.2. Review selection criteria and method

At the first stage, the relevant research outcomes were collected and sorted as part of the material selection process. A total of 67 published articles on LCA of renewable energy technologies were found. Most of these articles were published in the *Applied Energy Journal*, the *Renewable Energy Journal*, the *Journal of Cleaner Production*, the *International Journal of Life Cycle Assessment*, *Science of the Total Environment*, and the *Renewable & Sustainable Energy Reviews* journal. These articles were classified based on the type of renewable power resource and locations of the plants. Four renewable power technologies, i.e., solar-PV, wind, hydro, and biomass plants, were vastly studied through LCA.

In the next stage, the key LCA approaches were identified through an extensive review of the articles. The common system boundary for renewable energy technologies was also investigated. Moreover, the LCA mid-point and end-point impact categories were summarized from the published articles. Furthermore, LCA software and functional units were summarized reviewing the articles, which are highlighted in Section 2.3 of this chapter. In the following stage, the impactful elements and devices of the considered renewable plants based on resources and locations were analyzed and summarized.

In the last stage, the research gaps, limitations, and future recommendations were provided based on prior research studies. The reasons for the emissions from renewable energy plants and solutions to reduce the environmental impacts were suggested in this stage. LCA practitioners and sustainable energy producers will find the key materials/devices that need to be replaced for efficient LCA application and for cleaner production of renewable energy in the near future. Fig. 2.1 depicts a summary of the four main stages to conduct this review.

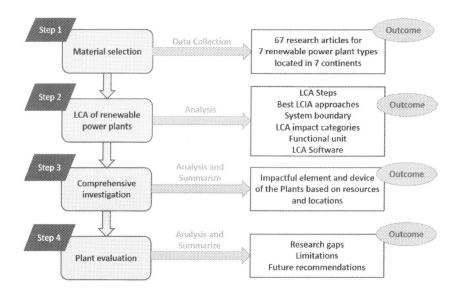

Figure 2.1 Systematic overview of the key steps followed to conduct this review.

2.3. Life cycle assessment of renewable power plants

This section highlights the steps, system boundary, and methods used for LCA analysis of renewable power plants.

2.3.1 Definition and steps of LCA

LCA is a well-recognized environmental impact evaluation tool which has been widely used to assess the effects caused by a unit or system on the environment [7]. This approach analyzes and classifies the impacts based on several standardized impact assessment mid-point and end-point indicators. The major LCA stages are depicted in Fig. 2.2. The key stages are resource acquisition, material processing, production, use and maintenance,

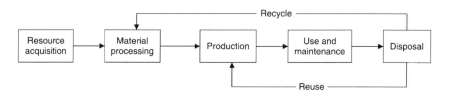

Figure 2.2 The key LCA stages [1].

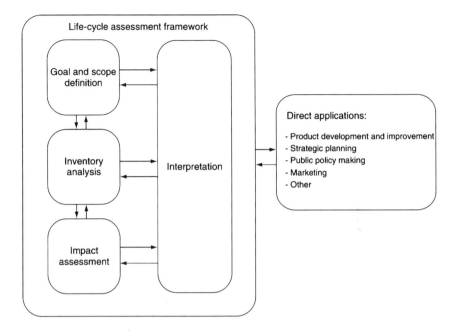

Figure 2.3 The LCA framework [1].

disposal, and recycling [1,59]. The LCA framework is depicted in Fig. 2.3. The key steps of LCA following the ISO 14040 are goal and scope definition, inventory analysis, impact assessment, and result interpretation. The applications of the LCA are strategic planning, public policymaking, and product development.

2.3.2 System boundary, functional unit, and software of LCA

The system boundary used in conducting LCA analysis is cradle-to-gate, cradle-to-grave, gate-to-gate, or gate-to-grave. Usually, the system boundary used in renewable energy-based LCA analysis is cradle-to-gate. A complete LCA system boundary in renewable energy technologies is comprised of inputs such as raw material extraction, transportation, water, organic and inorganic chemicals, and resources. The outputs of the system are electricity and waste emissions to soil, air, and water. The common life cycle inventory (LCI) for RESs is depicted in Fig. 2.4 [10]. The functional unit is considered as 1 kWh of electricity production. The most common software programs which are widely used in LCA of RESs are SimaPro and Gabi because of the integrated AusLCI database. The characterization and

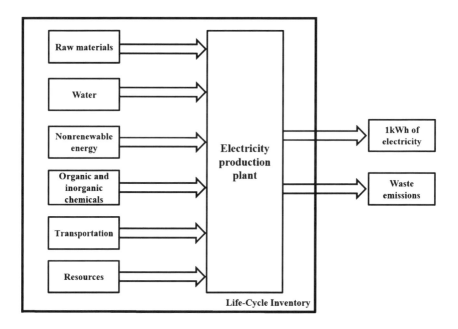

Figure 2.4 The common life cycle inventory for energy systems [2].

normalization of the LCA outcome is important as LCA impacts vary for the datasets of different companies, geographic locations, plant types, or analysis techniques.

2.3.3 Life cycle inventory datasets and geographic locations

The LCI datasets are usually gathered from various sources such as renewable or sustainable power industry reports, published studies, or renowned databases like ecoinvent and AusLCI. The datasets change from country to country. Therefore, global datasets are preferred for LCA analysis of renewable energy technologies. The accumulated datasets from the standard sources are quantified, aggregated, and validated for a specific renewable power plant. Most of the renewable power plant datasets are regional (for example, Asia, Europe, or America).

2.3.4 Life cycle assessment methods

The well-known and widely accepted LCA methods in renewable power technologies are the ReCiPe, Tool for the Reduction and Assessment of

State-of-the-art life cycle assessment methodologies applied in renewable energy systems   15

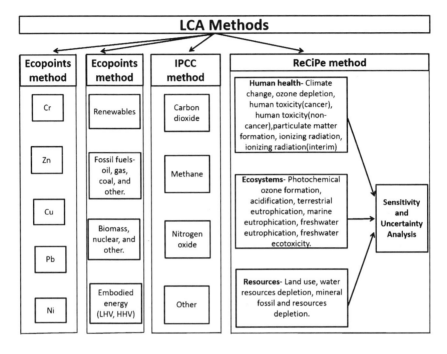

Figure 2.5 Schematic representation of LCA methods.

Chemical and other Environmental Impacts (TRACI), Institute of Environmental Sciences of the University of Leiden (CML), International Life Cycle Data System (ILCD), cumulative energy demand (CED), and Intergovernmental Panel on Climate Change (IPCC) approaches. These approaches differ from one to another based on characterization, normalization/weighting factors, and impact indicators. The choice of an appropriate method of LCA is dependent on the geography of the plant, characterization, and normalization factors. The LCA methods are schematically highlighted in Fig. 2.5.

The most commonly used problem-oriented mid-point indicators are carcinogens, noncarcinogens, respiratory inorganics, land occupation, ionizing radiation, aquatic ecotoxicity, terrestrial ecotoxicity, aquatic eutrophication, aquatic acidification, ozone layer depletion, and global warming. The most commonly used damage-oriented end-point indicators are climate change, human health, ecosystems, and resources. The elements used for quantification of environmental impacts and the best practice method for each impact indicators are depicted in Table 2.1 [7].

Table 2.1 The best practice method for each impact indicator [7].

Impact categories	Elements used for quantification	Best practice methods
Climate change	Quantification based on the human activities on climate based on greenhouse gas emissions. This is most commonly accounted for carbon dioxide, methane, and nitrous oxide emissions.	IPCC method to calculate GHG based on 100 years' emissions.
Resources depletion	Quantifies the depletion of natural resources from the earth, based on concentration of reserve and deaccumulation rate/quantity of fuels/ratio of annual production to available reserve/damage to resource based on increased cost of extraction.	CML method and ILCD method based on concentration of reserve and deaccumulation rate.
Eutrophication	Quantification based on the macronutrients released on the environment – air, water, soil. It can be aquatic and terrestrial. Aquatic eutrophication is quantified based on accelerated algae growth, reduced sunlight infiltration and oxygen depletion. Terrestrial eutrophication is quantified based on increase susceptibility of plants to diseases.	CML and IMPACT 2002+ which quantifies based on stoichiometric nutrification potential applicable to both categories.
Acidification	Quantification based on the acidifying impacts based on when acid precursor compounds are released onto the environment and deposited on land (terrestrial) or water (aquatic). Quantification is mainly based on nitrous oxides, sulfur oxides, sulfuric acid and ammonia.	CML and ILCD method based on critical load exceedance method of hazard index (HI) method.
Human toxicity and eco-toxicity	Quantifies the impact of toxicity substance released on land, water, and environment. Quantification is based on using pesticides, heavy metals, hormones, and organic chemicals.	USEtox, ILCD, ReCiPe, and IMPACT 2002+.

continued on next page

Table 2.1 (*continued*)

Impact categories	Elements used for quantification	Best practice methods
Photochemical ozone formation	Quantifies the impacts based on impacts from increase on ozone formation m troposphere. The main criteria are the emission of nitrous oxides, carbon monoxide, and those which impacts on ozone formation.	CML method, based on simplified description of atmospheric transport.
Particulate matter formation	Quantification is based on the emissions on air which are harmful for human health. In different LCIA methods, these are characterized under different impact categories.	TRACI as TRACI method distinguishes between different types of emissions.
Land use	Quantification is based on the amount of land use m LCA and its effects on biodiversity.	Currently no best practice methods for land use as no single method quantifies all levels of biodiversity.
Ozone depletion	Quantifies the impact based on the reduction m concentration of ozone.	Ozone depletion factors published by the World Meteorological Organizations (WMO).
Ionizing radiation	Quantifies the impact of radioactive species (radionuclides) on air and water.	ILCD or ReCiPe method based on the quantification of radioactive impact on human health.

2.4. LCA of renewable energy systems

This section discusses the LCA carried out on the key renewable power technologies, their analysis approaches, key findings, limitations, and future recommendations. The key renewable power plants include solar-PV, hydro, wind, and biomass plants.

2.4.1 Impact analysis of solar-PV power plants

Solar power is the most widely used renewable energy source all over the world. Table 2.2 describes the key analysis type, findings, and recommendations for LCA of solar-PV systems.

In total, 17 major LCA studies were conducted to assess the impacts of solar power plants. Dones et al. conducted LCA of PV systems considering Swiss studies on energy chains [60]. The analysis considered grid-connected PV plants for both monocrystalline and polycrystalline silicon cell technologies. The highest impact of the considered plants and location was due to high electricity requirements for the manufacture of plant elements. Tripanagnostopoulos et al. assessed the impacts of a solar-PV system and a hybrid solar-thermal system. The analysis considered a water-cooled photovoltaic thermal (PVT) system and assessed the impacts by the Eco-indicator method. The highest impacts of the solar plant were found to be due to PV modules, aluminum, and copper [61]. The authors claimed that the environmental profile could be improved by reducing the use of aluminum, water cooling, and diffuse reflectors in the system. Kannan et al. evaluated the effects of solar-PV systems in Singapore considering monocrystalline PV modules [62]. This analysis revealed that the environmental burdens from the solar-PV system are less than half of the burdens from the combined cycle plant. They recommended to replace the aluminum structure by a concrete structure to reduce energy use in PV module production, which results in a better environmental profile.

Moreover, a study carried out by Fthenakis et al. showed the effects of polycrystalline silicon, monocrystalline silicon, ribbon silicon, and thin-film cadmium telluride PV cells considering life cycles [63]. This analysis revealed that the release of waste products into the air could be reduced by 89% by replacing the conventional fossil fuel-based electricity generation systems in a grid with cadmium telluride PV systems of the same capacity. The authors suggested that the large incorporation of solar-PV-based energy in the grid could provide overall environmental benefits. Chow et al. evaluated the impacts of PV collectors with modular channel type aluminum absorbers [64]. Their analysis revealed that the vertical-mounted building-integrated PV system provided superior performance compared to the rooftop system. Fu and his research group carried out LCA and sensitivity analysis of the electricity and steam consumption during processing of poly-Si solar cells [65]. Their analysis revealed that the transformation of metallic silicon to solar silicon due to high electricity consumption is the most critical stage of the life cycle of the PV system in terms of environment. The LCA conducted by Yu et al. on flat-roof-mounted PV arrays showed that the considered solar-PV system is superior from all environmental perspectives [66]. They claimed that the use of hazardous materials in manufacturing the PV panels must be reduced and that more care is needed in end-of-life treatment. Wu et al. assessed the impacts of the ground-mounted solar power station and found that the energy payback

Table 2.2 Key findings and recommendations from recent studies on LCA of solar-PV plants.

Topic	System type/ analysis type	Key findings	Recommendation	Ref.
LCA of PV systems: results of Swiss studies on energy chains	Mono- and polycrystalline silicon cell technologies.	The most environmental impacts of the considered PV systems originate from high electricity requirements for manufacturing of its elements.	The environmental performance of future PV systems is expected to augment considerably owing to lower energy needs for manufacturing.	[60]
Energy, cost, and LCA results of PV and hybrid PVT solar systems	Multicrystalline silicon PV module and water-cooled PVT system.	The largest contributions to the overall impacts originate from PV modules, aluminum, and copper.	The heat recovery by using liquid cooling and diffuse reflectors provides better environmental performance.	[61]
LCA of solar-PV systems: An example of a 2.7 kWp distributed solar-PV system in Singapore	Monocrystalline solar-PV system.	The most GHG emissions (about 217 g CO_2/kW) to the environment come from the energy used for the manufacturing of PV modules.	Replacing the aluminum structure by a concrete structure results in less energy use in PV module production.	[62]
Emissions from PV life cycles	Multicrystalline silicon, monocrystalline silicon, ribbon silicon, and thin film cadmium telluride.	Thin film cadmium telluride PV releases the smallest amount of harmful GHG gases due to the lowest amount of energy consumption.	The environmental profiles of PV systems are improving due to enhanced material utilization rates.	[63]

continued on next page

Table 2.2 (*continued*)

Topic	System type/ analysis type	Key findings	Recommendation	Ref.
Environmental life cycle analysis of hybrid solar-PVT systems for use in Hong Kong	PV collectors with modular channel type aluminum absorbers.	The vertical-mounted building-integrated PV system provides superior performance compared with the rooftop system.	The used solar-thermal system demonstrates environmental superiority over other renewable options.	[64]
LCA of multi-crystalline PV systems in China	Sensitivity analysis is carried out to identify the impact of electricity and steam consumption during processing of poly-Si solar cells.	The highest critical stage of the life cycle of the PV system in terms of environment is found to be the transformation of metallic silicon into solar silicon.	The installation of multi-Si PV systems is recommended due to their better environmental performance.	[65]
Solar-PV development in Australia – a life cycle sustainability assessment study	Flat-roof-mounted PV solar array.	The large-capacity PV systems are more sustainable.	The use of hazardous materials in manufacturing the PV panels must be reduced.	[66]
Review on LCA of energy payback of solar-PV systems and a case study	Ground-mounted solar power station.	EPBT is found to be 2.3 years as the total energy input is 19.5548×10^6 MJ and the annual energy output is estimated to be 8.328×10^6 MJ.	The reduction of primary carbon-based energy usage in manufacturing the PV modules is recommended to significantly reduce environmental hazards.	[67]

continued on next page

Table 2.2 (*continued*)

Topic	System type/ analysis type	Key findings	Recommendation	Ref.
Environmental LCA of a rooftop solar panel in Bangkok, Thailand	Rooftop PV solar installation.	Key contributions to impacts are made by certain manufacturing stages and PV panel elements.	The recycling of materials is recommended to mitigate impact.	[68]
LCA of solar energy conversion systems in energetic retrofitted buildings	Sensitivity comparison between a solar-PV system and an equivalent conventional fossil fuel-based energy system.	The assessed cases depict that a retrofitted building can support 100% of its energy needs and cease emission of hazardous substances in less than 2 years.	The on-site electricity consumption paired with a net-metering policy scheme is suggested to incentivize installation of solar plants.	[69]
Prospects of LCA of renewable energy from solar-PV technologies: a review	Thin film, dye-sensitized, perovskite, and quantum dot-sensitized solar cells.	Monocrystalline silicon PV technology has the highest energy consumption, the longest EPBT, and the highest GHG emission rates compared with other solar-PV technologies.	The key step to reduce the GHG emissions of a solar-PV system is reduction of primary conventional energy consumption in producing the PV modules.	[70]
Comparative LCA of PV electricity generation in Singapore by multi-crystalline silicon technologies	PV systems with various p-type multi-crystalline silicon technologies.	The GHG emission rates of three roof-integrated PV systems with different p-type multi-Si PV technologies are in the range of 20.9 to 30.2 g CO_2-eq./kWh.	The shift from aluminum back surface field (Al-BSF) cell technology to passivated emitter and rear contact (PERC) cell technology will reduce the GHG emissions.	[48]

continued on next page

Table 2.2 (*continued*)

Topic	System type/ analysis type	Key findings	Recommendation	Ref.
Assessment of the geographical distribution of PV generation in China for a low-carbon electricity transition	Sensitivity analysis of solar-PV systems to return energy and carbon investments.	Distributed solar-PV systems are suitable for installation in Shandong and Jiangsu, China	Some policy implementations are needed for the future development of solar-PV technologies.	[71]
Comparative energy and GHG assessment of industrial rooftop-integrated PV and solar-thermal collectors	Sensitivity comparison between solar-thermal and solar-PV technologies.	The energy and GHG payback time ranges between 1.2 and 15 years and between 2 and 17 years, respectively.	In places where the solar-thermal system surpasses the solar-PV system, industries should use solar-thermal installations for heat production to supply process heat.	[72]
Life cycle and economic assessment of a solar panel array applied to a short-route ferry	Sensitivity assessment of solar power systems and diesel engine systems.	The longer payback time is associated with lower energy efficiencies and higher investment costs.	The minimal expense of the carbon credit would be $190/ton in making the marine business favorable.	[73]

time (EPBT) is 2.3 years as the total energy input is 19.5548×10^6 MJ and the annual energy output is estimated to be 8.328×10^6 MJ [67]. The authors concluded that there will be 27.7 years of net energy saving by operating a solar power station for 30 years instead of conventional power plants. Eskew et al. considered a rooftop solar-PV installation in Thailand and assessed impacts across all effect indicators [68]. They found that the GHG emission rate was 0.079 kg CO_2-eq./kWh over the lifespan of the considered framework and EPBT was estimated to be 2.5 years. These authors also recommended domestic manufacture of components and recycling of materials in order to reduce environmental effects in all categories.

Furthermore, Martinopoulos et al. conducted a comparative sensitivity analysis between a solar-PV system and an equivalent conventional fossil fuel-based energy system applied in a house [69]. This analysis suggested that the on-site electricity consumption paired with a net-metering policy scheme is environmentally superior for solar plant installations. Ludin et al. [70] showed that the monocrystalline Si PV systems have higher energy usage, EPBT, and GHG emission rates in comparison with other solar-PV systems. Luo et al. [48] reported that the GHG emission rates of three roof-integrated PV systems with different p-type multi-Si PV technologies were in the range of 20.9 to 30.2 g CO_2-eq./kWh. Of the considered PV systems, the long-term PV panels exhibited higher reliability with respect to environmental performance. The research conducted by Liu et al. [71] highlighted that the distributed solar-PV systems are suitable for installation in Shandong and Jiangsu, China, whereas these are not suitable for installation in Qinghai and Ningxia, China. Mousa et al. [72] compared the energy profiles and GHG emissions of industrial rooftop-integrated solar-PV and solar-thermal collectors. Their analysis revealed that the monocrystalline PV has a smaller environmental impact than polycrystalline PV. In places where the solar-thermal system surpasses the solar-PV system, industries should use solar-thermal installations for heat production to supply process heat. Moreover, in locations where the PV system surpasses the thermal system, industries should choose PV installations coupled with heat pumps to supply process heat. Recently, Wang et al. conducted LCA and sensitivity assessment of solar power systems and diesel engine systems [73]. Their analysis revealed that the minimal expense of the carbon credit would be $190/ton or more in making the solar panel array-based marine business favorable to a short-route ferry.

2.4.2 Impact analysis of hydropower plants

Hydropower is usually considered as the cleanest form of renewable electricity. However, although vast amounts of hydropower are produced across the globe, little research has been conducted to assess the hazardous emissions from hydropower plants by LCA analysis. Table 2.3 describes the analysis types, key findings, and recommendations from recent studies related to LCA of hydropower plants. Nine research groups have conducted LCA of hydropower plants. Pascale et al. evaluated the impacts of a community hydroelectric power system in rural Thailand using LCA analysis

[74]. This study revealed that smaller hydropower systems have a higher environmental impact per kWh than larger systems. Pang et al. assessed the environmental impacts of a small hydropower plant in China and showed that the major impacts are caused by the construction stage in the life cycle of the plant [75]. The authors claimed that the identification of suitable installation capacity and the use of modern equipment is essential to ensure the optimal output in terms of environmental performance of small hydropower plants in China. Kadiyala et al. [76] evaluated the life cycle GHG emissions from hydroelectricity generation systems and found that large-size diversion hydropower plants emitted the lowest amounts of life cycle GHGs (21.05 g CO_2-eq./kWh) among all considered cases. Li et al. [15] evaluated the carbon footprints of two large hydroprojects in China and showed that the reservoir is responsible for most CO_2 release and the sediment is responsible for most CH_4 release in the phase of dam decommission. Mao et al. conducted a sensitivity analysis of hydropower plants to investigate the sustainable development path [77]. They recommended an improved technology for an appropriate type of cooling system for hydropower plants to provide economic emancipation. Mahmud et al. [2] carried out an environmental sustainability assessment of a hydropower plant in Europe using LCA and found that the effects of hydropower plants in nonalpine zones on climate change are 10 times as strong as the effects of alpine plants. Ueda et al. assessed the environmental impacts of the construction stage of 11 microhydropower installations in the UK [78]. This analysis showed that the used metals in the hydroelectricity project contributed 86–98% of human toxicity and 79–98% of abiotic resource depletion impacts. The authors propose to replace the concrete construction of the hydropower plant with a wood-frame powerhouse as this reduces the impact on global warming by 6–12%.

2.4.3 Impact analysis of wind power plants

Wind power plants are a popular choice for renewable energy generation due to their higher reliability, cost-effectiveness, and pollution-free power generation. However, wind power plants mostly affect the environment in the stages of raw material extraction, parts manufacturing, structure building, and end-of-life recycling processes. Table 2.4 describes the analysis types, key findings, and recommendations from recent studies related to LCA of wind power plants. Seven research groups have conducted LCA of wind power plants. Turconi et al. conducted LCA of wind electricity

Table 2.3 Key findings and recommendations from recent studies on LCA of hydropower plants.

Topic	System type/ analysis type	Key findings	Recommendation	Ref.
LCA of a community hydroelectric power system in rural Thailand	Sustainability assessment of a community hydropower plant considering the construction, operation, and end-of-life stages.	The hydropower plant yields better environmental and financial outcomes than a diesel generator.	The hydropower plant is recommended for community electrification to meet environmental challenges.	[74]
Environmental LCA of a small hydropower plant in China	LCA-based impact assessment of a hydropower plant in China based on several categories.	The major impacts are made during the construction stage in the life cycle of the plant.	Structural design optimization, quality construction material, and good production practices are suggested.	[75]
Evaluation of the life cycle GHG emissions from hydroelectricity generation systems	The life cycle GHG emissions of various hydropower system types are quantified.	Large-size diversion hydropower plants emit the smallest amounts of GHGs (21.05 g CO_2-eq./kWh).	The site-specific characteristics of the hydroelectricity system need to be comprehensively considered.	[76]
Carbon footprints of two large hydroprojects in China: LCA according to ISO/TS 14067	Sustainability analysis of hydropower plants considering the reservoir operation, maintenance, and dam decommission stages.	The reservoir is responsible for most CO_2 release and the sediment is responsible for most CH_4 release during the phase of dam decommission.	The dam decommission stage is indicated as the most sensitive phase for releasing maximum GHG emissions.	[15]

continued on next page

Table 2.3 (*continued*)

Topic	System type/ analysis type	Key findings	Recommendation	Ref.
The sustainable future of hydropower: A critical analysis of cooling units based on the theory of inventive problem solving and LCA methods	Sensitivity analysis of hydropower plants to investigate the sustainable development path.	The evaporative cooling generator is indicated as a better option with lower GHG emission rates (2.59×10^{-1} kg CO_2-eq./kWh) than the air cooling generator (6.04×10^{-1} kg CO_2-eq./kWh).	Improved technology for an appropriate type of cooling system is recommended.	[77]
Environmental sustainability assessment of a hydropower plant in Europe using LCA	Environmental burdens of alpine and nonalpine hydropower plants are analyzed.	The hydropower plants of alpine areas offer environmentally superior performance with respect to global warming.	The most risky materials in plant installation must be identified and replaced with superior alternatives.	[2]
LCA of the construction phase of 11 micro-hydropower installations in the UK	The environmental hazards during the construction phase of 11 small hydropower plants are identified.	The use of upstream concrete contributes 25–44% to the global warming impact indicator.	It is suggested to replace the concrete construction of the hydropower plant with a wood-frame powerhouse.	[78]

production systems and compared them with other renewable and nonrenewable electricity generation systems in terms of environmental effects [38]. Under high wind penetration and low cycling conditions of the plant, it is recommended to increase the storage capacity to reduce operational costs. The emissions are reduced when the wind power plant is coupled with base load coal. The research carried out by Huang et al. on LCA of offshore wind power technologies showed that the use of ferrous metal affects human health and ecosystems [79]. This analysis also revealed that the use of fuels and synthetic materials mostly affects resources. The waste materials are suggested to be recycled to increase environmental performance. Chipindula [40] and his research team assessed the lifetime environmental impact of onshore and offshore wind plants in Texas and found that the GHG payback times and EPBT are in the range of 6–14 and 6–17 months, respectively, for the onshore farms. They recommended the replacement of lignite coal with wind to improve mid-point categories. Steel recycling could provide a 20% reduction of the average impacts of 15 mid-point impact indicators. Xu et al. [41] considered onshore wind power systems in China and conducted LCA analysis to quantify their environmental performances. A significant decrease in acidification, eutrophication, human toxicity, and ecotoxicity was found for wind plants compared to coal- and natural gas-based power plants. Manufacture of the tower contributes to global warming, and manufacture of the rotors contributes to abiotic depletion. It is suggested to optimize the structural design and the usage of raw materials to improve the environmental profiles of wind power generation systems in China. Schreiber et al. [42] carried out a comparative LCA of electricity generation by different wind turbine types. Their findings show that the use of copper, steel, and rare earth permanent magnets are the key factors for the environmental impacts of wind power plants. The research by Siddiqui [56] and his team in Canada revealed that the wind power plant contributes more to acidification, eutrophication, photochemical ozone creation, and human toxicity compared to other plants. Sustainable construction of wind turbines is suggested to improve the environmental profiles of wind power plants. Fang et al. [43] carried out a research to assess the impacts of a wind power–hydrogen coupled integrated energy system and indicated that the material production step is the strongest contributor of environmental effects for both hydro- and wind power plants.

2.4.4 Impact analysis of biomass power plants

Biomass power plants are providing a substantial share in the global electricity market. A small number of LCA analyses have been conducted to make this technology more sustainable and make electricity production cleaner. Table 2.5 describes the analysis types, key findings, and recommendations from recent studies related to LCA of biomass power plants. Beagle et al. carried out a comparative LCA of biomass application for electricity production in the EU and the US [44]. This analysis depicted that the utilization of biomass for electricity production reduces life cycle GHG release

Table 2.4 Key findings and recommendations from recent studies on LCA of wind power plants.

Topic	System type/ analysis type	Key findings	Recommendation	Ref.
LCA of electricity generation technologies: overview, comparability, and limitations	The cycling emissions during part-load operation and start-ups by a wind power plant in Ireland are estimated by the LCA approach.	The cycling emissions are responsible for about 7% of CO_2, NO_x, and SO_2 emissions in the considered plant.	The emissions are reduced when the wind power plant is coupled with base load coal.	[38]
LCA and net energy analysis of offshore wind power systems	Life cycle environmental impacts of an offshore wind power plant.	Waste recycling results in a 25% smaller environmental effect.	It is recommended to reduce the use of steel and concrete when building wind turbines.	[79]
Life cycle environmental impact of onshore and offshore wind farms in Texas	Wind power plants located in onshore, shallow water and deep water of Texas and the Gulf coast.	Raw material extraction is the major phase causing most environmental burdens.	The replacement of lignite coal with wind is recommended.	[40]
LCA of onshore wind power systems in China	Environmental performance is calculated for utility-scale wind power systems in China.	Manufacture of the tower contributes to global warming and manufacture of the rotors contributes to abiotic depletion.	Structural design and usage of raw materials are recommended to improve the environmental profiles of wind power systems in China.	[41]

continued on next page

Table 2.4 (continued)

Topic	System type/ analysis type	Key findings	Recommenda-tion	Ref.
Comparative LCA of electricity generation by different wind turbine types	The environmental effects of onshore wind turbines in Germany are evaluated.	The strongest impacts are associated with the manufacture of fundament, tower, and nacelle, which account for about 19%, 30%, and 99% of single impacts, respectively.	Recycling of waste is recommended for optimal improvement.	[42]
Comparative assessment of the environmental impacts of nuclear, wind, and hydroelectric power plants in Ontario: an LCA	Environmental hazards of nuclear, wind, and hydropower plants in Ontario are assessed and compared.	Wind power plants contribute more to acidification and human toxicity than other plants.	Sustainable construction of wind turbines is suggested.	[56]
Life cycle cost assessment of wind power–hydrogen coupled integrated energy system	Environmental effects of solar-PV, wind, and hydropower plants are analyzed.	The impacts of a wind power plant are higher than the impacts of hydro and solar-PV plants.	This study provides a helpful tool for the upcoming environmental impact evaluation of energy systems.	[43]

compared to the coal baseline. The policy implications are recommended in the perspective of present US and EU policies. Biomass-related factors such as transportation type and distance are suggested to be considered carefully to decrease GHG release. LCA and sensitivity analysis of electricity generation through biomass technologies in Portugal conducted by Loucao et al. showed that inputs can considerably affect the final environmental outcome, together with the quality of the biomass [45]. It is suggested that the best

Table 2.5 Key findings and recommendations from recent studies on LCA of biomass power plants.

Topic	System type/ analysis type	Key findings	Recommendation	Ref.
Comparative LCA of biomass utilization for electricity generation in the EU and the US	Impacts of biomass transportation on GHG emissions are assessed.	The utilization of biomass for electricity production reduces life cycle GHG release compared to the coal baseline.	The policy implications are recommended in the perspective of present US and EU policies.	[44]
Life cycle and decision analysis of electricity production from biomass – a case study in Portugal	An LCA model of biomass energy technology is developed and sensitivity analysis is carried out.	Cofiring using forest residues requires less energy consumption for the production of 1 MJ electricity.	The best ways are the use of manure and municipal waste for feedstock and residue gasification.	[45]
LCA of forest-based biomass for bioenergy: a case study in British Columbia, Canada	LCA is applied to four combustion- and gasification-based biomass energy technologies operated in British Columbia.	Avoidance of combustion of residues and fossil fuels gives a better environmental performance.	Improvements in energy conversion efficiency and disposal of wood ash are recommended.	[46]

ways are the use of manure and municipal waste for feedstock and residue gasification. Maier and his group conducted an LCA analysis of forest-based biomass for bioenergy generation and indicated that avoidance of fossil fuel combustion results in better environmental performance, especially with respect to the indicators of acidification, eutrophication, fossil resource depletion, respiratory effects, and photochemical ozone formation [46]. The energy conversion efficiency and the disposal of wood ash are the most influencing factors in bioenergy supply chains. Uncontrolled burning creates the highest environmental burden by destroying forest biomass chains, and hence must be controlled.

2.4.5 Impact analysis of other renewable power plants

Table 2.6 describes the analysis types, key findings, and recommendations from recent studies related to LCA of other power plants such as power generation using hydrogen energy carriers, electricity cogeneration from sugarcane bagasse, and biogas production and utilization substituting for grid electricity. Camacho et al. assessed the life cycle environmental impacts of biogas production and utilization substituting for grid electricity [80]. This analysis showed that the amount of GHG savings ranges from 524 to 477 kg CO_2-eq./MWh for biogas as a fuel by substituting fossil petrol and diesel fuels. The study by Rosenbaum et al. on LCA of torrefied and nontorrefied briquette use for heat and electricity generation showed a considerable reduction in the impact on global warming through replacing carbon-based fuels with the considered bioproducts owing to the use of propane in their processes [81]. To improve the environmental profile, field-dried feedstock with low moisture content is recommended. Ozawa and his team carried out a study on the life cycle CO_2 emissions from power generation using hydrogen energy carriers and showed that H_2 and NH_3 monoring energy production can have 52% and 36% lower life cycle CO_2 emissions than natural gas combined cycle energy production [82]. Ozturk et al. carried out LCA analysis of hydrogen-based electricity generation in place of conventional fuels for residential buildings and found that the electricity production with hydrogen is the most environmentally sustainable for most impact categories [83]. A recent analysis conducted by Lisperguer et al. on the sustainability of energy coproduction from sugarcane in Jamaica depicted that it added economic, environmental, and social value [84].

2.5. Geographic location-wise LCA of renewable energy systems

This section discusses the geographic location-wise LCA carried out on key renewable power technologies, their analysis approaches, key findings, limitations, and future recommendations. The considered locations of the renewable plants are Asia, Europe, America, and Australia.

2.5.1 Impact analysis of plants in Asia

Different countries in Asia, such as Indonesia, China, Singapore, and Pakistan, have shown extensive incorporation of renewable energies in the

Table 2.6 Key findings and recommendations from recent studies on LCA of other renewable plants.

Topic	System type/ analysis type	Key findings	Recommendation	Ref.
Life cycle environmental impacts of biogas production and utilization substituting for grid electricity, natural gas grid, and transport fuels	The environmental effects of biogas production concentrating on GHG emissions are assessed.	The amount of GHG savings ranges from 524 to 477 kg CO_2-eq./MWh for biogas as a fuel by substituting fossil petrol and diesel fuels.	It is recommended to identify the limitations of LCA for future improvements of bioenergy systems.	[80]
Life cycle impact and exergy-based resource use assessment of torrefied and nontorrefied briquette use for heat and electricity generation	The cradle-to-grave environmental impact of two bioproducts (nontorrefied and torrefied briquettes) is investigated and resource use is assessed.	The outcome shows a considerable reduction in the impact on global warming through replacing carbon-based fuels.	To improve the environmental profile, field-dried feedstock with low moisture content is recommended.	[81]
Life cycle CO_2 emissions from power generation using hydrogen energy carriers	LCA analysis is carried out to evaluate GHG emissions from hydrogen monofiring and ammonia monofiring power plants.	H_2 and NH_3 monofiring energy production can have 52% and 36% lower life cycle CO_2 emissions than natural gas combined cycle energy production.	It is recommended to extend the LCI to remaining low-carbon supply chains such as hydrogen energy carriers from locally existing resources.	[82]

continued on next page

Table 2.6 (*continued*)

Topic	System type/ analysis type	Key findings	Recommendation	Ref.
LCA of hydrogen-based electricity generation in place of conventional fuels for residential buildings	The environmental impacts of electricity production from hydrogen systems for residential purposes are assessed.	Electricity production with hydrogen is the most environmentally sustainable case for most impact categories.	Impacts on global warming can be reduced by adopting proper policies.	[83]
Sustainability assessment of electricity cogeneration from sugarcane bagasse in Jamaica	This study provides a few important messages to energy policy-makers to produce electricity with sugarcane cogeneration.	A complete environmental LCA analysis of the sugarcane- and biomass-based electricity power plant is carried out.	It is recommended to produce electricity by cogeneration from sugarcane to add environmental value.	[84]

main electricity grid. Researchers have assessed the environmental impact of electricity mixes at these locations by LCA analysis. Table 2.7 describes the analysis types, key findings, and recommendations from recent studies related to LCA of renewable electricity mixes in Asia.

You et al. conducted an economic and GHG savings analysis of decentralized biomass gasification to provide electricity to the rural areas of Indonesia and suggested a hybrid LCA to assess the impacts for various biomass type power plants [47]. The research group of Li et al. in China analyzed the carbon footprints of two large hydroprojects and suggested that the reservoir-based hydroelectricity plants are more sustainable due to their reduced environmental hazards [15]. Luo and his research team [48] conducted a comparative LCA of solar-PV-based electricity production in Singapore by multicrystalline Si technologies. This analysis showed that PV modules with frameless double glass can reduce the environmental effects

Table 2.7 Key findings and recommendations from recent studies on LCA of renewable plants in Asia.

Country	Topic	Plant type	Key findings	Recommendation	Ref.
Indonesia	Techno-economic and GHG savings assessment of decentralized biomass gasification for electrifying the rural areas of Indonesia	Biomass	The global warming effects of a biomass gasification plant are evaluated using the LCA approach.	A hybrid LCA is suggested to assess the impacts of various biomass type power plants.	[47]
China	Carbon footprints of two large hydroprojects in China: LCA according to ISO/TS 14067	Hydro	A unique LCA system boundary for two hydropower generation plants is developed.	Reservoir-based hydroelectricity plants are suggested for their reduced environmental hazards.	[15]
Singapore	A comparative LCA of PV electricity generation in Singapore by multi-crystalline silicon technologies	Solar-PV	The environmental impacts of three roof-integrated solar-PV technologies in Singapore are assessed.	The PV modules of silicon are suggested by various case studies.	[48]
Pakistan	Life cycle sustainability assessment of electricity generation in Pakistan: policy regime for a sustainable energy mix	Solar-PV, wind, hydro, coal, and gas	The most sustainable renewable power plant with the smallest environmental and economic impacts is the hydropower plant.	Gas-based power generation systems should be avoided in future energy mixes as they have a negative effect on the environment.	[49]

continued on next page

Table 2.7 (*continued*)

Country	Topic	Plant type	Key findings	Recommendation	Ref.
Singapore	Environmental impacts of transitioning to renewable electricity in Singapore and the surrounding region: an LCA	Solar-PV, hydro, geo-thermal	The shift toward renewables from fossil fuels will diminish global warming effects, but the risks related to acidification and human toxicity might increase.	The ideal renewable mix can provide greater sustainability from an environmental perspective.	[50]

compared to alternative options. Akber et al. of Pakistan carried out a life cycle sustainability assessment of electricity generation systems and depicted that the most sustainable renewable power plant with the smallest environmental impact is the hydropower plant [49]. Quek et al. performed an LCA analysis to assess the environmental effects of transitioning to renewable energy for Singapore and the nearby areas [50]. This analysis indicated that the shift toward renewables from fossil fuels will diminish global warming effects, but the risks related to acidification and human toxicity might increase.

2.5.2 Impact analysis of plants in Europe

Different countries in Europe, such as France, Italy, Germany, Portugal, and Greece, have vastly integrated renewable energies in the main electricity grid. Researchers have assessed the environmental impacts of electricity mixes at these locations by LCA analysis. Table 2.8 describes the analysis types, key findings, and recommendations from recent studies related to LCA of renewable electricity mixes in Europe.

Perilhon et al. compared the impacts of 2-MW and 10-MW biomass plants in France through LCA analysis. The evaluation of the impacts associated with nonregulated pollutants is suggested to be considered in future research [51]. Ardente and his team in Italy conducted LCA analysis of a solar-thermal collector and suggested to use a global database for quantifying the impacts [52]. Nugent et al. of Germany assessed the life cycle

Table 2.8 Key findings and recommendations from recent studies on LCA of renewable plants in Europe.

Country	Topic	Plant type	Key findings	Recommendation	Ref.
France	LCA of electricity generation from renewable biomass	Biomass	A comparison between the impacts of 2-MW and 10-MW biomass plants is depicted through LCA analysis.	The evaluation of the impacts associated with nonregulated pollutants is suggested to be considered in future research.	[51]
Italy	LCA of a solar-thermal collector	Solar-thermal	The environmental profiles of solar-thermal collectors to fulfill the sanitary warm water demand are investigated.	It is recommended to use a global data source. It is mainly applicable for the solar systems in Italy.	[52]
Germany	Assessing the life cycle GHG emissions from solar-PV and wind energy: a critical metasurvey.	Solar-PV and wind	The physical features of solar systems are mostly responsible for harmful emissions.	A procedure to lower GHG emissions is suggested to improve sustainability of the plants.	[53]
Portugal	Life cycle sustainability assessment of key electricity generation systems in Portugal	Solar-PV, wind, hydro, coal, natural gas	Hydropower is found as the best option from environmental and socioeconomic perspectives.	Renewable investments and resources that should be utilized to improve energy generation sustainability are recommended.	[54]

continued on next page

Table 2.8 (continued)

Country	Topic	Plant type	Key findings	Recommendation	Ref.
Greece	Integrated life cycle sustainability assessment of the Greek interconnected electricity system	Solar-PV, wind, hydro, biogas, and lignite	The solar-PV plant is found to be the most sustainable form to generate electricity for social purposes. The wind power plant is found to be the most sustainable form from environmental and economic perspectives.	The environmental impacts, costs, and social aspects are suggested to be adjusted choosing the most sustainable alternatives.	[55]

GHG emissions from solar-PV and wind energy plants. It was found that the physical features of the solar systems are most accountable for harmful emissions [53]. Kabayo and his group conducted the life cycle sustainability assessment of major energy production technologies in Portugal and found that hydropower is the best option from environmental and socioeconomic perspectives [54]. Roinioti et al. conducted an integrated life cycle sustainability assessment of the Greek interconnected electricity system [55]. Their analysis indicated that wind power is the most sustainable form of electricity from environmental and economic perspectives.

2.5.3 Impact analysis of plants in America

Different countries in America, such as Canada, Ecuador, and Brazil, have shown extensive incorporation of renewable energies in the main electricity grid. Researchers have assessed the environmental impacts of electricity mixes in the US by LCA analysis. Table 2.9 describes the analysis types, key findings, and recommendations from recent studies related to LCA of renewable electricity mixes in America.

Siddiqui et al. carried out a comparative LCA analysis among nuclear, wind, and hydroelectric power technologies in Canada [56]. These authors report that the impacts associated with the upstream and decommission-

Table 2.9 Key findings and recommendations from recent studies on LCA of renewable plants in America.

Country	Topic	Plant type	Key findings	Recommendation	Ref.
Canada	Comparative assessment of the environmental impacts of nuclear, wind, and hydroelectric power plants in Ontario: a life cycle assessment	Nuclear, wind, and hydro	Wind power plants are mostly responsible for environmental hazards due to acidification, eutrophication, and human toxicity.	The impacts associated with the upstream and decommissioning phases of the wind electricity generation systems need to be regulated.	[56]
Ecuador	Analysis of GHG net reservoir emissions of hydropower plants in Ecuador	Hydro	The GHG releases of two reservoir-based hydro-electricity plants are assessed and compared.	It is recommended to consider the social perspectives in LCA analysis.	[36]
Brazil	LCI for hydroelectric generation: a Brazilian case study	Hydro	A unique LCI is built for LCA of the hydropower plants in Brazil to assess the environmental hazards.	It is suggested to consider all the environmental effect indicators in a comprehensive LCA analysis.	[57]
Brazil	Performance analysis of a PVT solar collector in a tropical monsoon climate city in Brazil	Solar-thermal	This work highlights the most impacting parameters of a solar-thermal plant that harms the environment.	It is recommended to estimate GHG release in a step-by-step manner.	[31]

ing phases of the wind electricity generation systems need to be regulated as these are identified as the main impacting stages of the plant. Hidrovo et al. analyzed the GHG net reservoir emissions of hydropower plants in Ecuador [36]. This analysis also considered all possible options of release in their lifetime. Ribeiro et al. [57] and Rubio et al. [31] quantified the performance of a solar collector in Brazil. Their work highlights the key impacting parameters of a solar-thermal plant that harms the environment. It is suggested to consider all environmental effect indicators in a comprehensive LCA analysis.

2.5.4 Impact analysis of plants in other zones

Researchers at other geographic locations like Australia and South Africa have assessed the environmental impacts of renewable electricity mixes by LCA analysis. Table 2.10 describes the analysis types, key findings, and recommendations from recent studies related to LCA of renewable electricity mixes in Australia and South Africa.

Mahmud et al. analyzed the environmental impacts of solar-PV and solar-thermal technologies by LCA [10]. The environmental impacts of a

Table 2.10 Key findings and recommendations from recent studies on LCA of renewable plants in other zones.

Country	Topic	Plant type	Key findings	Recommendation	Ref.
Australia	LCA of environmental impacts of solar-PV and solar-thermal systems	Solar-PV and solar-thermal	The environmental impacts of a solar-PV plant are higher than those of a solar-thermal system of the same capacity.	It is recommended to conduct sensitivity analysis for all elements of both solar plant types.	[10]
South Africa	LCA of fossil carbon dioxide emissions in coal biomass-based electricity production	Biomass	The land use and occupation impacts of the biomass combustion processes are assessed.	The impacts of land transformation are recommended to be estimated to improve sustainability.	[85]

solar-PV plant are found to be higher than the impacts of a solar-thermal system of the same capacity. It is recommended to conduct sensitivity analysis for all elements of both solar-PV and solar-thermal plants. Benetto et al. conducted LCA analysis of carbon dioxide release in coal biomass-based energy generation [85]. The land use and occupation impacts of the biomass combustion processes are assessed in this study. The impacts of land transformation are recommended to be estimated to improve sustainability.

2.6. Summary and outlook

This section discusses the key LCA findings of renewable energy technologies. Analysis results are discussed based on five key LCA approaches utilized for evaluating the effects of the renewable electricity generation systems, major impacting elements of each plant at different geographic locations, fossil fuel consumption rates by the plants in their life cycle, key mid-point impact indicators of the considered renewable electricity generation systems, and key damage indicators of the renewable plants in their life cycle.

2.6.1 Choice of LCA methods to be used in LCA of renewable power plants

The choice of methods for assessing the impacts of renewable power plants under different categories is summarized as follows:
- The IPCC method is the preferred choice by most research groups for assessing the GHG emissions of renewable power plants based on a 100-year time frame.
- The CED approach is the best option for assessing the carbon-based energy consumption rates over the lifetime of renewable power plants.
- The Australian indicator method is used for assessing the impacts of plants in Australia.
- The TRACI method is mostly used for LCA studies of renewable power plants in America.
- The CML method is mostly used for LCA analysis of renewable power plants in Europe.
- In renewable power plants, the most used method for end-point damage assessment is the ReCiPe method.
- The ILCD method is commonly used for mid-point impact assessment of renewable plants under several effect indicators.

- The IMPACT 2002+ method is the best practice method for assessing the impacts of renewable power plants which quantifies aquatic and terrestrial eutrophication.

2.6.2 Major impacting elements of each plant at different geographic locations

The main impacting elements of renewable power plants depend on the type of renewable resources, geography, and the location of the plant. For the solar-PV plant, the PV modules are the most impacting element, whereas heat storage causes the maximum impact for solar-thermal energy systems. The wind turbines are the most impacting element for wind energy systems. The construction phase is mostly responsible for the impacts by a hydropower plant. Research carried out by Siddiqui et al. showed that the construction and decommissioning stages are key steps to affect the environment [56]. Hanafi et al. reported that the maximum impact by the hydropower plant in Indonesia is on marine aquatic ecotoxicity and abiotic depletion [86]. Li et al. showed that the reservoir and dam of the hydropower plant are mostly responsible for the impacts in China [15]. LCA analysis of hydropower plants in Thailand by Pascale et al. revealed that the global warming potential rate is 52.7 g CO_2-eq./kW [74]. LCA of a hydropower plant in Brazil by Geller et al. indicated that the global warming potential is 5.46 kg CO_2-eq./MW [87]. Overall, transportation and construction are key phases for the release of GHGs by wind, hydro, and biomass power plants. Differences in environmental impact between renewable plants at different geographic locations are due to variation in resources, raw materials, transportation types, and distances.

2.6.3 Power plant life cycle fossil fuel consumption rates

The rates of fossil fuel-based energy consumption by solar-PV, wind, hydro, and biomass plants were obtained from previous important works and are summarized here. Solar-PV systems, which are smaller than solar-thermal plants of the same capacity, use fossil fuels like coal and gas in their life cycle. On the other hand, biomass-based energy plants mostly consume oil-based fossil fuels in their life cycle. Wind power plants consume gas-based fossil fuels and nuclear power in their life cycle. Hydropower plants mostly consume coal-based fossil fuels. These fossil fuel consumptions in the life cycles of renewable power plants have a negative impact on the environment. Therefore, it is recommended to abate carbon-based energy

Table 2.11 Comparison of key mid-point impacts among various renewable plants.

Label	Solar-PV	Hydro	Wind	Biomass
Carcinogens	0.0129	0.007	0.2352	0.2667
Noncarcinogens	0.006	0.0003	0.0078	0.9293
Respiratory inorganics	0.0004	0.0002	0.0039	8.9191
Ionizing radiation	0.069	0.4668	11.1324	0.1915
Ozone layer depletion	0.0002	0.0022	0.0517	0.1296
Respiratory organics	2.11E−06	5.73E−05	0.0013	100
Aquatic ecotoxicity	0.0483	3.81E−05	0.0009	1.097
Terrestrial ecotoxicity	1.1698	0.0013	0.045	0.1248
Terrestrial acid	0.0006	0.0001	0.0035	14.796
Land occupation	0.0695	0.0054	0.1252	0.1906
Aquatic acidification	0.0002	0.0001	0.0027	6.3693
Aquatic eutrophication	0.1788	0.0074	0.1735	0.4919
Global warming	0.0006	0.0024	0.0149	3.6549

by replacing it with alternative renewable energy sources in the lifetime of the plants.

2.6.4 Comparison of key mid-point impacts among renewable electricity generation systems

Based on previous studies, the highest environmental impacts of solar-PV, wind, hydro, and biomass plants are caused by carcinogens, noncarcinogens, ionizing radiation, ozone layer depletion, terrestrial ecotoxicity, land occupation, aquatic eutrophication, and global warming. The results depicted in Table 2.11 and Fig. 2.6 show that the maximum impact of wind power plants lies in the ionizing radiation category and the strongest effect of biomass power plants lies in the global warming category. Solar plants are mostly responsible for terrestrial ecotoxicity and hydropower plants have smaller impacts in all mid-point categories.

2.6.5 Comparison of key damage caused by renewable plants in their life cycle

Based on previous LCA studies, the key damage to human health, the climate, and ecosystems caused by solar-PV, wind, hydro, and biomass plants considering cradle-to-grave effect outcomes is summarized in Figs. 2.7 to 2.10. The damage to ecosystems by solar-PV plants is maximal, with a rate of 99%. However, the impacts of wind power plants on the climate

State-of-the-art life cycle assessment methodologies applied in renewable energy systems   43

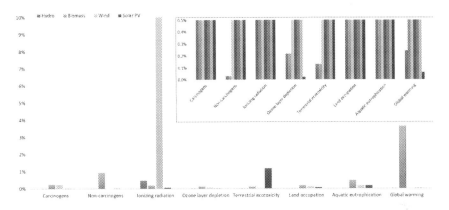

Figure 2.6 Comparison of key impacts of various renewable plants [2].

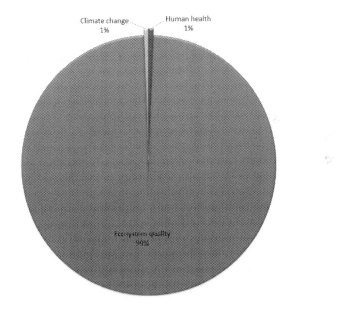

Figure 2.7 Key impacts of solar-PV plants.

are highest, with a rate of 52%. The impacts of biomass plants on human health are highest, with a rate of 47% [2,10]. To reduce these damages to the environment and to augment clean renewable power production, research should be carried out to find the most impacting materials in plants and replace them with equivalent or superior alternatives.

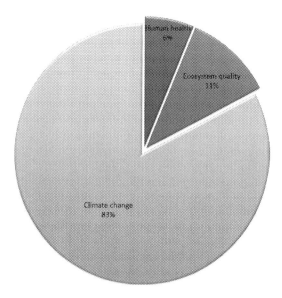

Figure 2.8 Key impacts of hydropower plants.

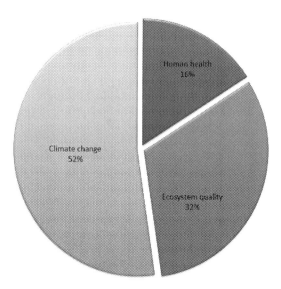

Figure 2.9 Key impact comparison with wind power plants.

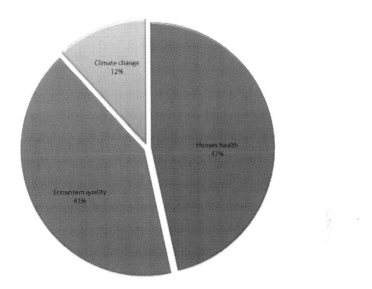

Figure 2.10 Key impact comparison with biomass power plants.

2.7. Conclusion and future recommendation

This chapter presents a comprehensive review of all available studies on the life cycle environmental impact assessment of renewable energy technologies such as solar–PV, wind, biomass, and hydropower plants at various geographic locations. Published works are compiled, analyzed, and compared, research gaps for future consideration are identified, and recommendations are made. The most impacting elements, devices, and stages of the considered renewable power plants are identified to replace them by sustainable alternatives. Therefore, the main contribution of this review is sixfold. First, it highlights the LCA methods of RESs and indicates the issues that must be overcome. Second, it analyzes the key findings and recommendations for the sustainable development of RESs in recent studies for future improvements. Third, it summarizes the key impacting elements of each plant in different countries to replace them by equivalent sustainable alternatives. Fourth, it compares the rates of fossil fuel-based energy consumption during the overall lifespan of the RES. Fifth, it estimates the rates of GHG emission from various renewable power plants. Last, it analyzes and summarizes the impacts of RESs on human health, ecosystems, climate, and resources. According to the LCA analysis results of RESs depicted in this chapter, it is evident that for the future development of cleaner and

sustainable power plant systems, the following recommendations should be considered:
- Replacement of fossil fuel by renewables is of paramount importance. The selection of renewable resources will rely on the geographic position of the plant.
- The development of a global LCA method is essential for each renewable power plant which would be valid irrespective of the geographic area of the plant.
- The development of a database focusing on renewable energy technology-based LCI datasets is essential, which should consider data for all categories of RESs all over the world.
- It is recommended to identify the most hazardous materials in plants and substitute them with environmentally superior alternatives.
- Recycling of waste is recommended for optimal improvement.
- Heat recovery using liquid cooling and diffuse reflectors provides better environmental performance.
- In places where the solar-thermal system surpasses the solar-PV system, industries should use solar-thermal installations for heat production to supply process heat.
- To reduce the GHG emissions of a solar-PV system, the reduction in primary conventional energy consumption in producing the PV modules is a key step.

CHAPTER THREE

Environmental impacts of solar-PV and solar-thermal plants

The demand for clean energy is high, and the shift from fossil fuel-based energy to environmentally friendly sources is the next step to eradicating greenhouse gas (GHG) emissions. Solar energy technology has been touted as one of the most promising sources for low-carbon, nonfossil fuel energy production. However, the true potential of solar-based technologies is established by augmenting efficiency through satisfactory environmental performance in relation to other renewable energy systems. This chapter presents an environmental life cycle assessment (LCA) of a solar-photovoltaic (PV) system and a solar-thermal system. Single crystalline Si solar cells are considered for the solar-PV system and an evacuated glass tube collector is considered for the solar-thermal system in this analysis. A life cycle inventory (LCI) approach is developed considering all inputs and outputs to assess and compare the environmental impacts of both systems based on 16 impact indicators. LCA has been performed by the International Life Cycle Data System (ILCD), Impact 2002+, cumulative energy demand (CED), Eco-points 97, Eco-indicator 99, and Intergovernmental Panel on Climate Change (IPCC) methods using SimaPro software. The outcomes reveal that the solar-thermal framework provides about five times higher release into the air (100%) than the solar-PV framework (23.26%), and the outputs of the solar-PV system into soil (27.48%) and solid waste (35.15%) are almost three times lower than those of the solar-thermal system. The findings also depict that the solar panel is responsible for most impact in the considered systems. Moreover, uncertainty and sensitivity analyses have also been carried out for both frameworks, which reveal that the Li-ion batteries and the copper indium selenide (CIS)-solar collectors perform better than others for most of the considered impact categories. This study indicates that a superior environmental performance can be achieved by both systems through careful selection of the components, taking into account the toxicity aspects, and by minimizing the impacts related to the solar panel, the battery, and heat storage.

3.1. Introduction

The accelerating growth of the world economy and the exponential rise of the global population have resulted in an increasing demand for traditional fossil fuel-based power production, resulting in the emission of record levels of GHGs [88,89]. This burning of fossil fuels has significant environmental effects, such as air pollution, global warming, ozone layer depletion, acid rain, and climate change [90,91]. Therefore, researchers have been trying to develop alternative sustainable energy technologies to overcome the challenges of the energy crisis and reduce the environmental impact [92,93]. Due to the growing demand for renewable energy sources in the last decade, PV technologies have received considerable attention because of their high potential for large-scale sustainable energy generation with high efficiency and a superior environmental profile with low carbon dioxide (CO_2) emissions [94–97]. It is generally considered that solar technologies have smaller environmental effects than conventional power-generating units. However, it has not been explored which one is more environmentally friendly between solar-PV and solar-thermal systems, and the impacts from each element, including production, transportation, installation, operation, and end-of-life recycling, on the environment have not been compared.

The research goal of this project is to assess the environmental effects of solar-PV and solar-thermal frameworks by a systematic LCA approach and compare the findings to make better-informed choices. Significant advancements in the efficiency and economic viability of PV technologies like solar-PV systems and solar-thermal systems have been achieved by various research groups [98–100]. Several research studies have been conducted to calculate the effects of a single element of solar technologies like PV panels [22–25], batteries [18], and solar-thermal collectors [52]. Others carried out country-based LCA research [26–33] to assess the impacts of solar technologies; they have not considered a global database, which is required for the consistent global practice of LCA [101] and global policy development [29]. Moreover, separate LCA analyses have been conducted to identify the hazards of solar technologies that directly affect the human body [102–105], but the rate of energy use from fossil fuel-based sources in manufacturing solar-PV and solar-thermal systems has not been measured. In this study, a literature review of LCA-based analyses of solar-thermal and solar-PV systems was performed. The summary is provided in Tables 3.1 and 3.2, respectively. However, very few studies have assessed and compared the environmental impacts of the individual parts of both frameworks, which is required to identify the problematic elements and to replace them by equivalent environment-friendly options. Therefore, given the remarkable role that solar technologies play in reducing global GHG emissions, it is necessary to investigate the environmental profiles of solar-PV and thermal systems considering each element of the system to understand their full potential.

Herein, the ecological impacts of each of the elements of solar-PV and solar-thermal systems, like the solar collector, battery, converter, inverter, power meter, breaker, flow meter, valve, pump, heat transfer fluid (HTF) tank, etc., are assessed by LCA and compared. In the LCA analysis both systems are considered for a small residential area. The GHG emissions and fossil fuel-based energy consumption by each element of both frameworks are also estimated. Moreover, sensitivity analysis of the considered systems is conducted by varying the PV panels and the battery storages to determine the best option in terms of their effects on the environment. Therefore, the key contributions of this chapter can be summarized as follows:

- A comprehensive LCI is developed considering inputs and outputs of solar-PV and solar-thermal systems.
- The environmental impacts of elements of both frameworks are assessed based on a "cradle-to-grave" scheme with 16 impact indicators.
- The effects of both systems are compared.
- The GHG emissions from both systems throughout their lifetime are quantified.
- The amount of fossil fuel-based power consumed during the manufacturing of elements, installation and operation of the systems, and the waste management period is estimated.
- Sensitivity and uncertainty analyses are conducted for both systems.

The current study is different from some of the previously reported studies [90,108–110] and provides additional information on the impacts associated with each of the elements, like the PV panel, valve, battery, converter, controller, flow meter, etc., in both solar-PV and thermal systems, and the critical materials and stages are identified; it is anticipated that by replacing the hazardous materials the environment can be saved from long-term dangerous emissions.

In the light of the above, the rest of the chapter is organized as follows. The materials and methods are briefly described in **Section 3.2**. The re-

Table 3.1 Previous works of life cycle assessment of solar-thermal systems and their limitations.

Source Ref.	Topic	Main focus of the work	Limitations
[52]	Life cycle assessment of a solar thermal collector	The environmental performances of solar thermal collector for sanitary warm water demand has been studied.	Data source is not global. It is only applicable for the plants of Italy.
[31]	Performance of a PV thermal solar collector in a tropical monsoon climate city in Brazil	This work identifies the most important parameters that affect the efficiency of the collector when operating in a locality of tropical monsoon climate zone in Brazil.	The estimation of step by step GHG emission has not been undertaken.
[105]	Comparative experimental LCA of two commercial solar-thermal devices for domestic applications	The environmental analysis for Flat Plate Thermos-phonic Units (FPTU) and integrated collector storage (ICS) solar water heaters is carried out through an LCA study.	They have not considered the end-of-life phase in assessing the impacts.
[106]	Life cycle analysis of a solar-thermal system with thermochemical storage process	The LCA technique to study the environmental impacts associated with solar-thermal system (SOLARSTORE) is highlighted in this work. The raw material acquisition and components manufacturing processes contribute 99% to the total environmental impacts during the whole life cycle.	The installation and maintenance processes are excluded from LCA.

sults and discussions revealing the environmental performances of solar-PV and thermal systems are presented in **Section 3.3**. **Section 3.4** highlights

Table 3.2 Previous works of life cycle assessment of solar-PV system and their limitations.

Source Ref.	Topic	Main focus of the work	Limitations
[90]	Review of the life cycle GHG emissions from different PV and concentrating solar power electricity generation systems	The cadmium telluride PVs and solar pond CSPs contributed to minimum life cycle GHGs.	The environmental contributions from the solar chimney and solar pond electricity generation systems have not been revealed.
[96]	Life cycle analyses of organic PV: a review	The environmental impacts of organic PVs under several indicators like CED, EPBT, and GHG emissions are discussed.	Lack of the evolution of environmental sustainability assessment in case of organic PVs.
[24]	LCA of an innovative recycling process for crystalline silicon PV panels	An innovative process for the recycling of silicon PV panels was analyzed.	The end-of-life phase was neglected in this analysis.
[25]	LCA of PV panels: a review	Energy-related indicators such as EPBT and indicators related to climate change such as the CO_2 emission factor were examined.	The electronic properties of the panel or BOS components were not evaluated.
[107]	LCA of grid-connected PV power generation from crystalline silicon solar modules in China	For multi-Si, mono-Si, LS-PV, and distributed PV systems, PV station manufacturing is responsible for about 84% or even more of total energy consumption and GHG emissions.	The solution to cut down the energy consumption and GHG emission during PV station manufacturing was not discussed.

continued on next page

the limitations of this research. Finally, concluding remarks on the research outcomes and recommendations are highlighted in **Section 3.5**.

Table 3.2 (continued)

Source Ref.	Topic	Main focus of the work	Limitations
[102]	Perovskite PV: LCA of energy and environmental impacts	The results of Eco-indicator 99, the EPBT, and the CO_2 emission factor were compared among existing PV technologies, and further uncertainty analysis and sensitivity analysis were performed for the two modules.	The way to lower the CO_2 emission factor was not mentioned and this work lacks information to improve the system performance ratio and the device lifetime.
[53]	Assessing the life cycle GHG emissions from solar-PV and wind energy: a critical metasurvey	Physical characteristics of the solar technologies are mostly responsible for emissions.	The procedure to lower GHG emissions was not highlighted.

3.2. Materials and methods

3.2.1 Solar-PV and solar-thermal system overview

The solar-PV system consists of a monocrystalline PV panel, a DC-to-DC charge controller, a DC-to-AC inverter, a power meter, a breaker, and a battery (Fig. 3.1). The PV panel generates DC electricity which is controlled by the charge controller (DC/DC) to obtain a regulated DC output. The controlled DC current is then stored in the battery. The DC output from the battery is inverted to AC by the DC/AC inverter. A power meter is utilized to measure and record the flow of electricity.

Fig. 3.2 shows the schematic framework of the solar-thermal system. It consists of a solar collector, flow meter, pump, HTF tank, ball and check valve, heat storage, temperature gauge, boiler, and reservoir. The PV collector harvests solar energy and converts it into heat, which is utilized to warm the cold water. The temperature gauge and flow meter measure the water temperature and flow rate, respectively. The pump is used for circulating the water throughout the system. The heat is stored between the boiler, the HTF tank, and the solar collector. The boiler is a vessel where water is heated.

Environmental impacts of solar-PV and solar-thermal plants 53

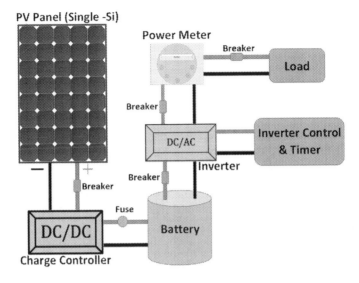

Figure 3.1 Schematic framework of the solar-PV system.

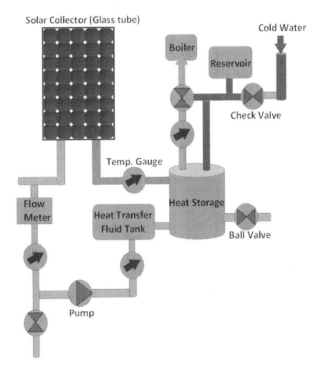

Figure 3.2 Schematic framework of the solar-thermal system.

3.2.2 Life cycle assessment method

The main aim of this research is to determine the environmental impacts of solar technologies like solar-PV and solar-thermal systems and compare their effects on the environment based on 16 impact indicators. For that reason, a well-ordered LCA method is used. LCA is a very effective approach for evaluating the environmental hazards of any device or system [111–113]. LCA has been widely used in impact estimation, sensitivity analysis, and sustainability testing [11,114–117]. This LCA has been accomplished maintaining International Standardization Organization (ISO) standards 14040:2006 and 14044:2006 [118,119]. In this study, SimaPro version 8.5 was used to evaluate the ecological threats of solar technologies. Herein, LCA is carried out by creating an LCI considering all of the elements for both solar technologies. The following four basic steps are maintained in LCA analysis:

1. goal and scope definition, where the LCA objective is highlighted and boundaries are determined following ISO 14040 [119];
2. LCI, where the energy-, material-, and emission-based input–output flows are assembled following ISO 14041 [118];
3. life cycle environmental impact evaluation, where impacts are assessed based on 16 effect indicators following ISO 14042 [118];
4. impact outcome interpretation, where obtained effects are annotated and examined with the objective of the LCA following ISO 14043 [119].

The major LCA steps are explained in the following sections to highlight the LCA approach carried out in this work.

3.2.2.1 Goal and scope definition

The first step of LCA is goal and scope definition. The main objective of this LCA is to assess and compare the ecological hazards of solar-PV and thermal systems. LCA is carried out considering both mid-point (cradle-to-gate) and end-point (cradle-to-grave) aspects for both frameworks [117,120]. Therefore, the total LCA takes into account all life cycle stages for both systems such as raw material extraction, key element production, transportation, framework installation, and waste management. The functional unit of the LCA is considered as 1 kWh of energy production, which determines the reference flow rates [111,113].

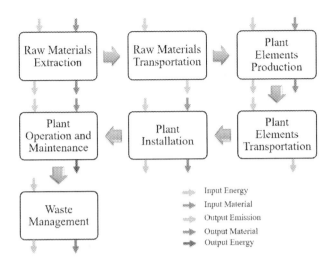

Figure 3.3 Step-by-step energy and material flows for both systems.

3.2.2.2 Life cycle inventory

The formation of the LCI is the second step of LCA. The input resources like raw materials and energies and the output emissions per unit process are considered in assembling the LCA inventory. Fig. 3.3 shows the energy and material flows for both systems in a step-by-step manner. Both solar technologies follow the same steps in their lifespan such as raw material extraction from mines using resources, raw material transportation to the plant location for manufacturing key materials, key material production at the plant, transportation of the produced materials to the solar plant area, installation and operation of the plant, and finally the end-of-life waste management. At each step there are input and output flows as marked in the figure. We created a comprehensive LCI for both frameworks. Fig. 3.4 shows the system boundary considered in this research. The ecoinvent database [121,122] is used to gather the input and output flows, as it has international industrial and commercial data for material production, transportation, energy consumption, etc. [101,121,122]. Table 3.3 shows the data source for each of the key elements of the frameworks. An assembly is formed using all the unit processes of the solar-PV system, which is then used to assess the individual impacts of each process element. Likewise, another assembly is built using all unit processes of the solar-thermal system to determine the effects of each element.

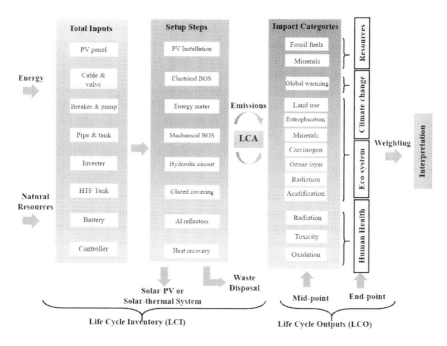

Figure 3.4 System boundary of the LCA.

3.2.2.3 Life cycle impact evaluation

Life cycle impact evaluation is the third step of LCA analysis. The rate of reference flow is considered for one functional unit (1 kWh). Impacts were assessed using SimaPro software for both frameworks. The ILCD method is utilized in evaluating the effects. This approach considers the inputs from raw material extraction to manufacturing, transportation, and usage (cradle-to-gate or mid-point) [102], and gives outputs for 16 impact indices. It gives potential environmental impacts under major effect indicators like global warming, climate change, land use, toxicity, and acidification. The cradle-to-grave (end-point) LCA analysis was carried out by the Impact 2002+ approach [123] based on 14 impact categories: carcinogens, noncarcinogens, respiratory inorganics, ionizing radiation, ozone layer depletion, respiratory organics, aquatic ecotoxicity, terrestrial ecotoxicity, terrestrial acid, land occupation, aquatic acidification, aquatic eutrophication, global warming, and nonrenewable energy, which are further subdivided into four major indicators, namely human health, ecosystem quality, climate change, and resources. The Raw Material Flows (RMF) method tracks the mass flow of all inputs and outputs based on adding

Table 3.3 Data collection for frameworks in the solar-PV system and the solar-thermal system.

Assembly	Unit process	Process source
Solar-PV system	Converter	Converter, 500 W, for electric system {GLO}/production/Conseq, U
	Battery	Battery, Li-ion, rechargeable, prismatic {GLO}/market for/Alloc Def, U
	Cable	Cable, unspecified, {GLO}/market for/Conseq, U
	Breaker	Switch, toggle, type, at plant/GLO U/AusSD U
	Power meter	Electric meter, unspecified {GLO}/production/Conseq, U
	Inverter	Inverter, 500 W, at plant/GLO U/I U/AusSD U
	PV panel	Photovoltaic panel, single-Si, at plant/GLO U/I U/AusSD U
Solar-thermal system	Reservoir	Tap water, at user/RER U/AusSD U/Link U
	Valve	Exhaust air valve, {GLO}/production/Conseq, U
	Boiler	Hot water tank, 200 L, at plant/GLO U/I U/AusSD U
	Solar collector	Solar collector glass tube, with silver mirror, at plant/GLO U/AusSD U
	Heat storage	Heat storage, 200 L, at plant/GLO U/I U/AusSD U
	Temp. gauge	Temperature, 500°F gauge {GLO}/market for/Conseq, U
	Pump	Pump, 40 W, at plant/GLO U/I U/AusSD U
	HTF tank	Expansion vessel, 200 L, at plant/GLO U/I U/AusSD U
	Flow meter	Flow meter, vortex type, at plant/GLO U/AusSD U

all the elementary flows available in the ecoinvent 2.0 database [124]. This RMF method is utilized to find and compare the input materials and output emissions of the systems. The CED approach is employed to calculate fossil fuel-based energy usage amounts for both considered frameworks, as this approach has been extensively used in assessing different sorts of fuel intakes throughout the lifetime of a unit [125]. CED considers various fuel inputs such as fossil fuels, renewables, nuclear, biomass, and embodied energy for the overall lifespan of both systems [126,127]. It is important to realize the consumption of carbon-based fuels to replace them by renewable sources for better environmental performance. Moreover, the IPCC approach is employed to assess the GHG emission rate. The IPCC approach reveals the climate change factors by the considered systems with a time frame of 100 years. This method usually considers hazardous gas emissions like carbon dioxide, methane, and nitrous oxide [128]. Uncertainty analysis has been conducted by the ILCD method to check the sustainability for the environmental impact indicators of the LCA analysis and to investigate the probability distribution of both systems following the method described in [97,102]. Finally, sensitivity analysis is conducted using the ILCD method considering different PV panels and battery storages to examine their effects.

3.2.2.4 Life cycle impact interpretation

Life cycle impact interpretation is the final stage of LCA. The impacts are assessed, compared, and interpreted in terms of the main factors responsible for environmental effects of solar-PV and thermal systems. Moreover, logical judgments are given based on uncertainty and sensitivity analysis.

3.3. Results and discussion

3.3.1 Environmental profiles of solar-PV systems

The overall inputs from nature and outputs into soil, water, and air and solid waste for each element of the solar-PV system are assessed by the RMF approach [124]. This approach gives output considering the raw material- and emission-based mass flow by adding all elementary flows. The life cycle inputs and outputs of the elements of a solar-PV system such as a PV panel, inverter, power meter, breaker, cable, battery, and converter are evaluated considering the energy and material flows in a step-by-step manner by the RMF method.

The obtained results are presented in Fig. 3.5 for comparison. Amounts are obtained in kilograms and the total amount of each element, like the converter, battery, cable, etc., is set at 100%. In case of the PV panel, it is clear from the figure that it intakes about 20% from nature during production and releases equally to air, soil, and solid waste (about 30%), but there is no direct release into water throughout the lifetime of the PV panel. On the contrary, the converter, battery, cable, and power meter of the solar-PV

Environmental impacts of solar-PV and solar-thermal plants 59

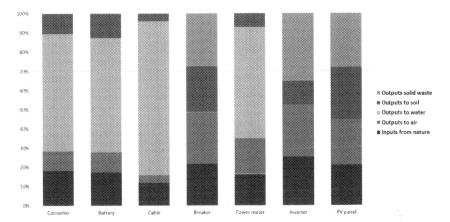

Figure 3.5 Life cycle inputs and outputs of the solar-PV system using the RMF methodology.

framework release a higher output to water, whilst the inverter, breaker, and PV panel release more output as solid waste (landfill). Therefore, the end-of-life recycling of these key parts is essential to overcome the problems associated with their release into water and soil.

The environmental impacts caused by each element of the solar-PV framework are depicted in Fig. 3.6, found by the ILCD method [102]. The maximum effects were observed for the climate change, ozone depletion, human toxicity, photochemical ozone depletion, acidification, terrestrial and marine eutrophication, and water resource depletion impact categories by the PV panel. On the other hand, the highest impacts from the battery were observed on mineral, fossil, and renewable resource depletion, land use, and freshwater eutrophication. As it is not possible to build a battery without chemicals, and these chemicals are mostly responsible for the harmful emissions and other effects of the battery, researchers should rethink the use of environmentally friendly materials without considering efficiencies. Furthermore, the power meter is mostly responsible for ionizing radiation because it carries much energy to ionize electrons from atoms during operation. This radiation is a risk for the human body as it can affect DNA and damage living cells. Therefore, future research should be directed toward developing an energy meter with the smallest ionizing radiation-based impact.

The results of the end-point LCA analysis of the solar-PV system using the Impact 2002+ approach [123] are depicted in Fig. 3.7. The PV panels are mostly responsible for affecting human health and climate change,

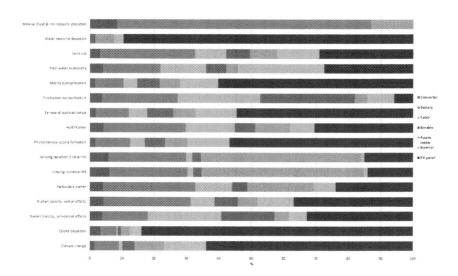

Figure 3.6 Environmental profiles of the considered solar-PV system.

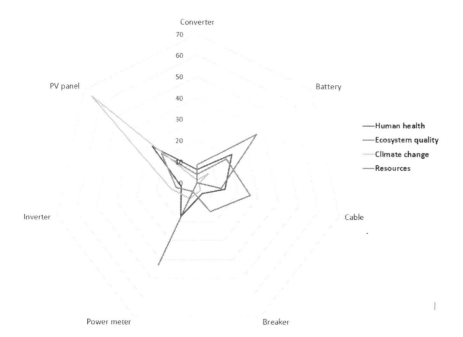

Figure 3.7 End-point impacts of the individual components of the solar-PV system.

whereas the battery mostly affects resources. The hazardous fluids used to transfer heat in solar modules are mostly responsible for the high impacts in categories like toxicity and acidification, which must still be overcome.

3.3.2 Environmental profiles of solar-thermal systems

In this section, the life cycle environmental hazards of a solar-thermal system are highlighted. The total input–output scenarios for each part of the solar-thermal system are evaluated by the RMF method [124]. The life cycle input–output rates of the solar-thermal framework are depicted in Fig. 3.8. The solar collector takes 25% of its total inputs from nature and others from other nonnature-based sources. It releases about 22% to solid waste, about 18% to the soil, about 7% to water, and about 18% to air. The boiler released mostly solid waste (about 48%). Almost 18% of its inputs comes from nature. However, the pump and valve of the solar-thermal system emit more output to water, while the boiler, reservoir, and temperature gauge emit significant outputs as solid waste. Moreover, the solar collector and heat storage release mostly to the air. However, the HTF tank and flow meter are totally discharged into water at the end-of-life. This happens because in the considered case these two parts are not recycled.

The cradle-to-gate (mid-point) environmental effects of each part of the solar-thermal system are highlighted in Fig. 3.9, obtained by the ILCD approach [102]. Among 16 impact types, the highest impacts from the solar collector are observed to climate change, ozone depletion, human toxicity, acidification, terrestrial eutrophication, ecotoxicity, water resource depletion, and land use. However, the maximum hazards from the valve are to freshwater eutrophication and mineral, fossil, and renewable resource depletion. The other parts of the framework show minor impacts. Overall, most of the impact from the solar-thermal system occur for the solar collector and heat storage (about 90% for the 14 impact categories) because of the use of hazardous materials and chemicals in their manufacturing, operation, and recycling.

The end-point LCA analysis of the solar-thermal system was conducted by the Impact 2002+ method [123]. The end-of-life results depicted in Fig. 3.10 reveal that the solar collector and the heat storage are the critical components in terms of the environment, as they are mostly responsible for damaging the ecosystem, climate, and human health. Moreover, the boiler

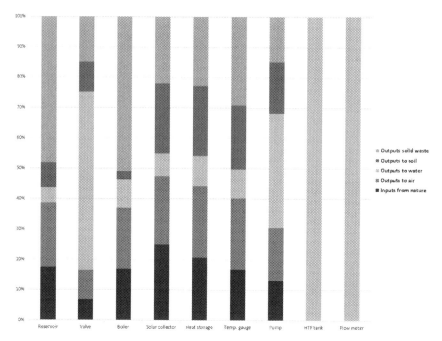

Figure 3.8 Life cycle inputs and outputs of the solar-thermal system using the RMF methodology.

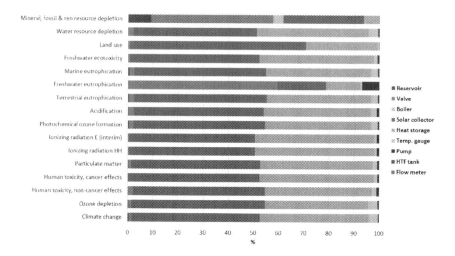

Figure 3.9 Environmental profiles of the considered solar-thermal system.

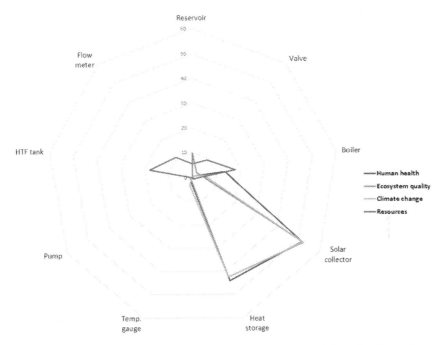

Figure 3.10 End-point impacts of the individual components of the solar-thermal system.

and HTF tank affect resources to a great extent in comparison with other parts of the framework.

3.3.3 Comparison of impacts between solar-PV systems and solar-thermal systems

The input–output comparison between the solar-PV system and the solar-thermal system is highlighted in Table 3.4, which shows that a higher input from nature is taken by the solar-thermal system than by the PV system. However, the outputs to the air and soil and solid waste are greater from the solar-thermal system than from the PV system, while outputs to water are mostly from the PV system.

The life cycle environmental impacts of the solar-PV and thermal systems as estimated by the ILCD method are depicted in Fig. 3.11. The results show that the solar-thermal system emits more hazardous materials and is highly responsible for more of the impact categories like land use, freshwater ecotoxicity, marine and terrestrial eutrophication, acidification,

Table 3.4 Life cycle inputs and outputs comparison between the solar-PV system and the solar-thermal system.

Label	Solar-PV system (%)	Solar-thermal system (%)
Inputs from nature	20.80	100
Outputs to air	23.26	100
Outputs to water	100	0.43
Outputs to soil	27.48	100
Solid waste outputs	35.15	100

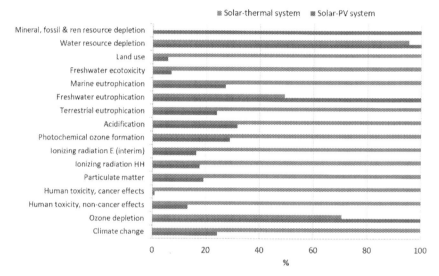

Figure 3.11 Comparison of environmental impacts from the solar-PV and the solar-thermal system.

photochemical ozone formation, ionizing radiation, particulate matter, human toxicity, and climate change than the solar-PV system.

The cradle-to-grave effect outcome comparison between the solar-PV and thermal frameworks is obtained by the Impact 2002+ method based on four major indicators, as demonstrated in Fig. 3.12. The outcome shows that a solar-thermal system is more dangerous for human health, the climate, and the ecosystem than an equivalent solar-PV system. However, the impacts on resources by the solar-PV system are 10 times as high as those by the solar-thermal framework.

Environmental impacts of solar-PV and solar-thermal plants 65

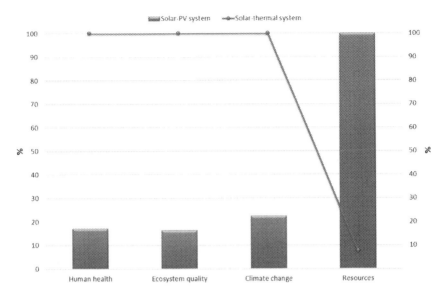

Figure 3.12 End-point impact comparison of the systems using Impact 2002+ methodology.

3.3.4 GHG emission factor estimation

The well-known IPCC method [128] is used to evaluate and compare the GHG emissions of the solar-PV and thermal systems. Figs. 3.13 and 3.14 show GHG releases by the solar-PV and thermal frameworks. Clearly, the maximum amounts of dangerous nitrous oxide and carbon dioxide emission are due to the cable and inverter, respectively. However, the converter, battery, and power meter are mostly responsible for land transformation. The valve, flow meter, and HTF tank from the solar-thermal system release most to the land. The solar collector, boiler, heat storage, and temperature gauge of the solar-thermal system emit higher amounts of nitrous oxide to the environment. The comparative GHG release obtained by the IPCC approach is demonstrated in Fig. 3.15, which shows that the solar-thermal framework releases about five times as much carbon dioxide as the solar-PV. Nitrous oxide emission is doubled for the solar-thermal system in comparison to the solar-PV system. However, other GHGs are emitted mostly by the solar-PV system. Overall, the obtained outcomes of this research confirm that cautious selection of less toxic solar panels, battery, and heat storage is a prerequisite to achieve a superior environmental performance by both systems.

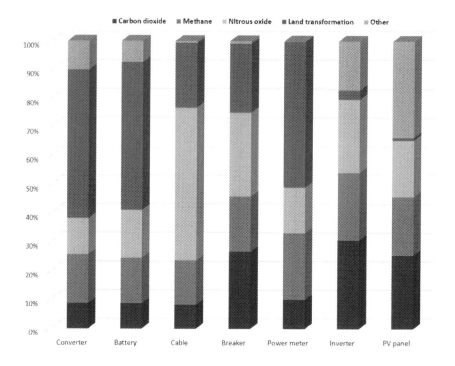

Figure 3.13 GHG emission of the solar-PV system with a time period of 100 years.

3.3.5 Fossil fuel-based energy consumption evaluation

The comparative amounts of fossil fuel-based energy consumption by the solar-PV and thermal systems as determined by the CED approach [125] of LCA are demonstrated in Fig. 3.16. The outcome clearly shows that solar-thermal installations consume more power than solar-PV systems. The gas-based fossil fuel consumption rate accounts for the maximum power usage by the solar-PV system. On the other hand, biomass-based energy accounts for the lowest amount of energy consumption by the solar-PV framework. Therefore, due to the smaller amount of fossil fuel consumption, the solar-PV installation is a better choice than the solar-thermal installation.

3.3.6 Sensitivity and uncertainty analyses

Two sensitivity analyses have been conducted to examine the environmental performance of the solar-thermal and solar-PV systems for different PV panels and batteries to identify the superior one in terms of the environmental impact. Table 3.5 shows the impacts of a solar-thermal framework

Environmental impacts of solar-PV and solar-thermal plants

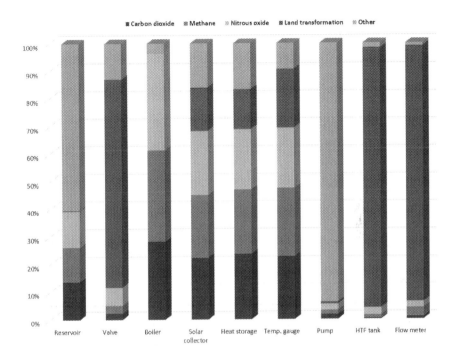

Figure 3.14 GHG emission of the solar-thermal system with a time period of 100 years.

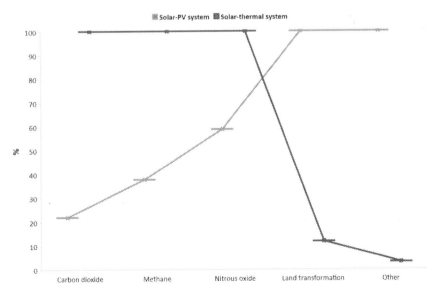

Figure 3.15 GHG emission of the systems as determined using IPCC methodology.

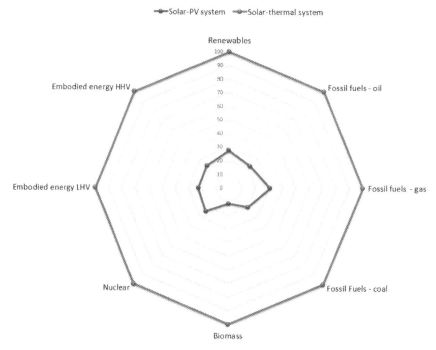

Figure 3.16 Required energy from different sources to build, operate, and dispose of both systems.

for five different solar collectors: amorphous silicon (a-Si), CIS, multi-Si, ribbon-Si, and single Si. The analysis outcome shows that single Si is highly responsible for climate change, whereas the CIS collector is the least hazardous to the climate. The solar collector made of a-Si is largely responsible for human toxicity and the multi-Si-based collectors are mostly responsible for water resource depletion. Overall, the solar collectors made from CIS have a superior environmental profile. Various types of batteries such as lithium-ion (Li-ion), sodium chloride (NaCl), and nickel–metal hydride (NiMH) are used in the sensitivity analysis of a solar-PV framework.

The analysis outcome based on different battery types for the solar-PV system highlighted in Table 3.6 shows that a NiMH-based framework provides higher impacts on indicators like acidification, particulate matter, ozone depletion, ionizing radiation, eutrophication, freshwater toxicity, water resource depletion, and climate change. The NaCl battery-based solar-PV system has a maximum impact in the human toxicity category. Overall, the Li-ion type battery-based solar-PV framework showed the best environmental profile. Therefore, stakeholders should consider CIS as a so-

Table 3.5 Sensitivity analysis outcome for different solar collector types for the solar-thermal system.

Impact category	a-Si [mPt]	CIS [mPt]	Multi-Si [mPt]	Ribbon-Si [mPt]	Single-Si [mPt]
Climate change	0.31	0.13	1.86	1.62	2.55
Ozone depletion	0.002	0.001	0.18	0.17	0.1837
Human toxicity, cancer effects	9.71	1.56	8.46	3.76	8.90
Particulate matter	0.11	0.03	0.61	0.51	0.707
Ionizing radiation HH	0.09	0.03	0.16	0.12	0.156
Ionizing radiation E	0.19	0.13	0.14	0.35	0.527
Photochemical ozone formation	0.10	0.04	1.01	0.92	1.28
Acidification	0.10	0.04	0.67	0.59	0.835
Terrestrial eutrophication	0.10	0.04	0.80	0.71	1.105
Freshwater eutrophication	0.001	0.002	0.09	0.09	0.094
Marine eutrophication	0.05	0.02	0.47	0.43	0.649
Freshwater ecotoxicity	0.41	0.09	1.68	1.50	1.767
Land use	9.9×10^{-5}	3.7×10^{-5}	5.1×10^{-5}	4.2×10^{-5}	5.9×10^{-5}
Water resource depletion	0.06	0.02	3.28	1.50	2.94
Mineral, fossil & resource depletion	5.16×10^{-11}	4.95×10^{-11}	1.23×10^{-8}	9.36×10^{-9}	1.2×10^{-8}

lar collector and a Li-ion battery as the energy storage device in building solar systems. The main implication of this result lies in the solar-PV and solar-thermal plant industry, where investors should use environmentally friendly parts to reduce the environmental impacts.

The probability distributions of the solar-PV and thermal systems are demonstrated in Figs. 3.17 and 3.18, which were obtained by using the Eco-indicator 99 approach of LCA. The bars with smaller rates depict

Table 3.6 Sensitivity analysis outcome based on different battery types for the solar-PV system.

Impact category	Li-ion [mPt]	NaCl [mPt]	NiMH [mPt]
Climate change	0.0559	0.0558	0.1813
Ozone depletion	0.0072	0.0053	0.5297
Human toxicity, noncancer effects	1.8839	1.3667	1.093
Human toxicity, cancer effects	1.0477	2.6478	2.2411
Particulate matter	0.123	0.455	1.045
Ionizing radiation HH	0.068	0.174	0.174
Ionizing radiation E (interim)	0.015	0.227	0.871
Photochemical ozone formation	0.040	0.092	0.208
Acidification	0.114	0.757	1.753
Terrestrial eutrophication	0.037	0.044	0.092
Freshwater eutrophication	0.058	0.045	0.054
Marine eutrophication	0.017	0.019	0.041
Freshwater ecotoxicity	0.258	0.260	0.383
Land use	0.00049	0.00051	0.00084
Water resource depletion	0.287	0.658	0.912
Mineral, fossil & resource depletion	0.806	0.670	2.111

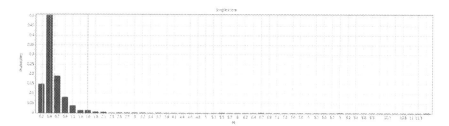

Figure 3.17 Probability distribution for the single-score impact category of the solar-PV system.

higher probabilities of getting identical impacts from the installations. The lower bars of the probability distributions for the single-score impact category of the solar-PV system with a rate of approximately 90% of the total probabilities indicate that this system is environmentally highly viable. Furthermore, about 70% of the small bars of the probability distributions for the single-score impact category of the solar-thermal system reveal that this system is also environmentally friendly and robust.

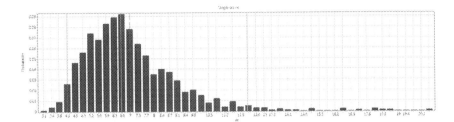

Figure 3.18 Probability distribution for the single-score impact category of the solar-thermal system.

3.4. Limitations of this study

The key limitations of this study are summarized as follows:
- LCA of solar technologies other than solar-PV and solar-thermal has not been studied in this research.
- LCA analysis of this study is completely dependent on the ecoinvent 2.0 global database.
- Sensitivity analysis has been conducted for the battery and the solar collector alone due to the lack of data sources.
- Replacement of elements that are responsible for hazardous emissions without considering the efficiencies and robustness of the systems has not been studied in this research.
- Ways to reduce the consumption of fossil fuels during the manufacturing of elements, installation, and operation of the systems and the waste management period have not been tracked in this work.
- Determination of the energy payback period and an economic estimation of the considered solar systems have not been accomplished.

Future studies should be directed toward overcoming the abovementioned limitations. Thus, solar systems can eliminate many long-term dangerous emissions.

3.5. Conclusions

In this chapter, the environmental effects of a solar-PV and a solar-thermal system are evaluated and compared through systematic LCA. To ensure the effectiveness of this research, (i) a comprehensive system boundary is developed for both of the considered solar technologies, (ii) LCA is carried out for both systems by multiple methods to assess the environmental profiles, (iii) the GHG emission rates are estimated for both systems, and

(iv) sensitivity and uncertainty analyses are conducted to examine the environmental performance of both systems more critically. The well-known SimaPro software and the renowned ecoinvent global database are used for assessing the life cycle environmental impacts by multiple methods such as ILCD for mid-point analysis, Impact 2002+ for end-point analysis, CED for fossil fuel-based energy consumption estimation, Eco-points 97 for metal- and gas-based emission assessment, Eco-indicator 99 for uncertainty analysis, and IPCC for GHG emission evaluation. The outcome of this research provides valuable information on the environmental impacts of each element of the considered solar technologies and can be used to identify more environmentally friendly options in for battery energy storage and solar collectors. The results highlight that the solar-PV framework is environmentally superior to the solar-thermal framework for most of the impact indicators, such as land use, freshwater ecotoxicity, marine and terrestrial eutrophication, acidification, photochemical ozone formation, ionizing radiation, particulate matter, human toxicity, and climate change. The outcomes also reveal that the solar-PV system is less impactful on human health, climate change, and ecosystems than the solar-thermal system. Sensitivity analysis shows that CIS-solar collectors and Li-ion batteries perform better than others for the solar-thermal and solar-PV systems, respectively. However, the main limitation of this work is that a sensitivity analysis for all considered elements has not been carried out. Thus, future work should include tracking the unsafe components and checking possible replacements to develop more environmentally friendly solar systems. Overall, it is recommended that, in order to protect the environment from hazardous emissions by the solar technologies, future research should be directed toward finding replacements for hazardous parts or processes that have a superior profile in terms of the environment.

CHAPTER FOUR

Environmental impacts of hydropower plants

Hydropower is a widely used source of clean energy which causes some hazardous emissions that affect human health, ecosystems, and resources. However, in spite of an enormous amount of hydropower generation in Europe, no research has been carried out to evaluate the hazardous emissions from plants located in alpine and nonalpine areas. Therefore, this chapter will analyze and compare the environmental impacts of hydropower plants in alpine and nonalpine areas of Europe by a systematic life cycle assessment (LCA) approach. The impacts are estimated by the ReCiPe 2016, Impact 2002+, and Eco-points 97 methods based on a number of effect indicators such as global warming, ozone formation, ecotoxicity, water consumption, acidification, eutrophication, ionizing radiation, carcinogenic radiation, ozone depletion, and land use. Moreover, the fossil fuel-based power consumptions and the greenhouse gas (GHG) emissions in the life cycle of hydropower plants in both locations are estimated using the cumulative energy demand (CED) and the Intergovernmental Panel on Climate Change (IPCC) methods, respectively. The outcomes reveal that hydropower plants of alpine regions offer a better environmental profile for the global warming indicator (2.97×10^{-5} kg CO_2-eq./MJ) than nonalpine plants (3.92×10^{-4} kg CO_2-eq./MJ), but the effects are nearly identical for the other indicators. Overall, the hydropower plants of nonalpine regions contribute 10 times more to climate change than alpine ones. The findings of this research will play a pivotal role in promoting sustainable production of hydropower, especially using the full potential of the alpine region, thus leading towards environmentally friendly clean renewable electricity generation.

4.1. Introduction

Currently the global energy demand is rising due to the growing world population and increasing industrial activity. The fossil fuel-based power generation rate has been increased to meet the growing demand [9,129]. These conventional power-generating units release dangerous GHGs during power production [30,100,130]. For that reason, the world climate is affected [10,11,28,97]. People are getting more concerned about ecosystems, thinking about the unavoidable threats that the hazardous emissions pose to the environment, and have agreed to reduce global CO_2 emissions to enhance the quality of life. Therefore, to save the environment and to reduce the use of conventional power generation systems, renewable energy production has become considerably more popular [38,47,93,95,131]. The coming decades will witness an unprecedented use of renewable sources for power production. Among renewable energy technologies, hydropower is considered to be the cleanest power generation source [12,13]. Hydropower plants have been installed in alpine and nonalpine regions of Europe due to the availability of resources [27,132,133]. However, previous research showed that hydropower plants release hazardous emissions during the construction phase [36,37]. Taking this fact into account, it is important to quantify the pollution caused by hydropower plants in alpine and nonalpine areas of Europe.

In spite of the enormous amount of hydropower generation all over the world, only a few studies have been carried out to evaluate the dangerous emissions during the life cycle of the hydropower plant [34–37]. A case study-based research conducted by Botelho et al. compared the environmental impacts of hydropower plants in Portugal, indicating that fauna and flora are most commonly affected [134]. They did not consider the whole life cycle of the plants. Ribeiro et al. examined the life cycle inventory (LCI) of hydroelectric power generation in Brazil, but they did not use a global database [57]. The carbon footprints of two large hydroprojects in China were investigated by Zhe Li et al. following LCA according to International Standardization Organization (ISO) 14067 [15]. They did not consider metal-based emissions and did not precisely quantify the impacts on human health, ecosystems, resources, and the climate. Pang et al. and Li et al. reported that the use of environmentally friendly materials and optimization of the plant design are required to reduce the impact of small hydropower plants in China, as they release most hazardous gas during the construction phase [15,75]. The rate of GHG emission and associated uncertainties in the LCA analysis of a hydropower plant in Japan were analyzed by Hondo et al. [135]. An Indian research group (Varun et al.) has considered three hydropower plants with different capacities and locations and assessed their impacts through life cycle analysis [58]. Hanafi et al. from Indonesia conducted LCA of a mini hydropower plant and suggested that carcinogenic and freshwater aquatic ecotoxicity during the construction phase are mostly responsible for the impact [86]. The impacts associated with a community hydropower plant in Thailand were assessed by Pascale et al. [74]. They showed that the considered plant performed better than a diesel generator from an environmental perspective. An investigation run by Geller et al. of a small hydropower plant in the Brazilian Amazon revealed that the steel used in the turbines and the concrete used in the buildings are crucial for the overall impact of the plant [87]. A Canadian research team (Siddiqui et al.) has compared the impacts of a hydropower plant with those of equivalent wind and nuclear plants [56]. According to their research, global warming, acidification, and eutrophication are the key impact indicators of a hydropower plant. The major contributions and relevant research gaps of these studies are highlighted in Table 4.1. However, to the authors' knowledge no research has been conducted to determine the impacts of alpine and nonalpine region-based hydropower plants and compare their effects. Therefore, it is essential to quantify the impacts by a systematic approach considering all input and output flows during the whole lifespan and the end-of-life waste management schemes of the plants.

This research is aimed at estimating the hazardous effects of hydropower plants of alpine and nonalpine areas by LCA, a widely accepted systematic approach to evaluate the ecological effects of a power production plant throughout its lifespan, from raw material extraction to processing, transport, operation, and end-of-life disposal [102,112,136,137]. SimaPro software version 8.5 is used in this research work to assess the impacts by systematic LCA [138]. The impacts are obtained based on 14 impact indicators: carcinogens, noncarcinogens, respiratory inorganics, ionizing radiation, ozone layer depletion, respiratory organics, aquatic ecotoxicity, terrestrial ecotoxicity, terrestrial acidification, land occupation, aquatic acidification, aquatic eutrophication, global warming, nonrenewable energy, and mineral extraction. The ReCiPe 2016 and Impact 2002+ methods are utilized to assess these effects. These effects are categorized under four major indicators: resources, climate change, ecosystem quality, and human health. The Eco-points 97 method is used to investigate metal- and gas-based dangerous emissions. Furthermore, emission of GHGs, such as carbon dioxide, methane, nitrous oxide, etc., into the air is estimated by the Intergovernmental Panel on Climate Change (IPCC) method. Therefore, the major contributions of this work can be summarized as follows:

- The environmental hazard of existing hydropower plants in Europe is estimated.

Table 4.1 Recent studies on LCA of hydropower plants and the research gaps.

Source Ref.	Topic	Major contributions	Research gaps
[57]	LCI for hydroelectric generation: a Brazilian case study	An LCI for hydropower plants in Brazil has been developed.	The estimation of environmental and social impacts has not been accomplished in this research.
[15]	Carbon footprints of two large hydro-projects in China: LCA according to ISO/TS 14067	Carbon footprints of the two projects are compared with over 150 worldwide cases.	The impacts of reservoir sediments of the plants have not been assessed.
[139]	A benchmark for life cycle air emissions and life cycle impact assessment of hydrokinetic energy extraction using LCA	GHG emissions have been quantified for a hydrokinetic energy extraction system and compared with other power generation types such as coal, gas, nuclear plants, etc.	The replacement of toxic fiberglass materials has not been identified, which must be replaced by a better alternative to improve the environmental profile of the considered system.
[140]	Ecosystem impacts of alpine water intake for hydropower: the challenge of sediment management	The impacts of flow abstraction and the challenges of sediment management upon ecosystems have been revealed through a systematic review. The main characteristics of natural alpine stream aquatic ecosystems have also been highlighted.	The impact of hydropower plants in the alpine zone has not been considered and explored in this work.
[141]	Impacts of climate change, policy, and the water–energy–food nexus on hydropower development	The impacts of policy, climate change, and the water–energy–food nexus on hydropower development have been reviewed at a global scale.	The social and environmental influences of hydropower generation systems have not been depicted.

continued on next page

Table 4.1 (*continued*)

Source Ref.	Topic	Major contributions	Research gaps
[128]	Emissions from tropical hydropower plants and the IPCC	This work points out that the emissions from dams need to be considered in inventories of the IPCC, which have been overlooked.	They have not considered other LCA approaches like ReCiPe, Impact 2002+, etc.
[142]	LCI of energy use and GHG emissions for two hydropower projects in China	The environmental performance of two hydropower plants in China has been assessed by an economic input–output-based LCA method.	The metal-based emissions have not been considered in this work.
[36]	Analysis of net GHG reservoir emissions of hydropower plants in Ecuador	The environmental impacts of dam- and run-of-river-based hydropower plants in Ecuador have been evaluated by the LCA approach.	The social aspects of the plants have not been assessed.
[143]	Addressing biogenic GHG emissions from hydropower plants by LCA	A statistical analysis of methane and CO_2 emissions from hydropower plants has been carried out to assess the sustainability.	The influences of the plants on human health, ecosystem quality, and climate change have not been highlighted.

- A unique LCI for hydropower plants located in both alpine and non-alpine areas of Europe is created to assess their environmental hazards in terms of ecosystems, climate change, resources, and human health.
- A step-by-step LCA analysis is performed to determine the GHG and metal-based emissions of both categories of plants.
- An uncertainty analysis for both categories of hydropower plants is conducted.

This research work is unique as it quantifies the environmental effects by several standardized approaches for the first time and considers both the mid-point and end-point impacts of both categories of plants considering the cradle-to-grave LCA boundaries.

In the light of the preceding, the rest of the chapter is organized as follows. **Section 4.2** highlights the countries of alpine and nonalpine areas in Europe and their hydropower production scenarios. In **Section 4.3** an overview of the LCA method is given. **Section 4.4** depicts the results of the environmental profiles, metal- and gas-based emissions, GHG releases, lifetime inputs and outputs, and energy consumption rates of the plants. **Section 4.5** discusses the impact outcome comparison with previous studies, impact comparison with other types of power plants, and uncertainty analysis of the plants. The limitations of this study are highlighted in **Section 4.6**. Finally, concluding remarks on the research outcomes and future recommendations are presented in **Section 4.7**.

4.2. Hydropower plants of alpine and nonalpine areas in Europe

The alpine region covers the countries which are surrounded by the Alps mountains. There are eight European countries which share their border with the Alps: Italy, France, Austria, Switzerland, Liechtenstein, Germany, Slovenia, and Monaco. Fig. 4.1 shows the boundary of the alpine countries in Europe [3]. Among these eight countries, five have significant production of hydropower. Fig. 4.2 shows the 25 European countries which have hydropower plants. Among these 25 countries, five countries are located in the alpine region, while the others are located in nonalpine regions. Fig. 4.2 also shows the installed capacity of hydropower plants per country with the number of installed plants. Details of the alpine countries with hydropower plants and their production details are shown in Table 4.2. The leading hydropower-generating countries in Europe are Germany, Austria, Italy, Spain, Sweden, and France [4]. Among these six countries, four are alpine countries with 42% of their total electricity production by hydropower plants. Table 4.3 shows the leading nonalpine countries with hydropower production. Tables 4.2 and 4.3 show the total numbers of hydropower plants in 2007 and 2011, their installed capacities, and their yearly production. According to the information presented in Table 4.2 for alpine countries, Germany is the leading country in hydropower plants installed in the alpine region, followed by Austria, Italy, France, and Slovenia. Table 4.3 illustrates the leading nonalpine countries with hydropower production: Sweden, the Czech Republic, Spain, and Poland [4].

Environmental impacts of hydropower plants

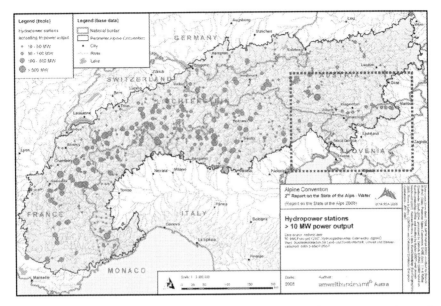

Figure 4.1 Map of the alpine boundary in Europe (source: 2nd Report on the State of the Alps) [3].

Table 4.2 Hydropower production details for the alpine areas in Europe.

Country	Plants in 2011	Plants in 2007	Capacity (MW in 2011)	Production (GWh/year)
Germany	7512	7503	1723	8352
Austria	2993	2354	1284	5778
Italy	2601	1835	2896	10,958
France	1935	1825	2110	6820
Slovenia	471	456	118	587

Table 4.3 Hydropower production details for the nonalpine areas in Europe.

Country	Plants in 2011	Plants in 2007	Capacity (MW in 2011)	Production (GWh/year)
Sweden	1867	1813	1283	4350
Czech Republic	1475	1405	297	1159
Spain	1250	553	1926	4719
Poland	739	681	281	1035

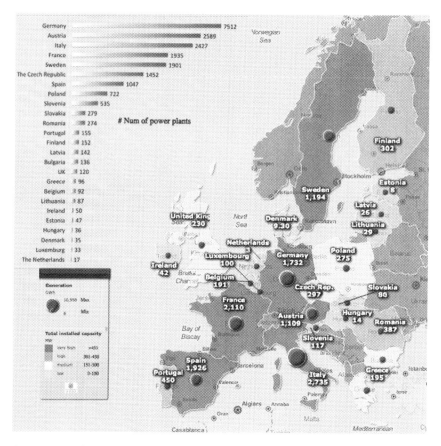

Figure 4.2 Hydropower production scenarios in alpine and nonalpine areas of Europe [4].

4.3. Methodology

The aim of this research work is to analyze the environmental impact of the hydropower plants of alpine and nonalpine areas of Europe. The analysis is conducted using the LCA approach, a powerful tool for checking the carbon footprints of any product, unit, or system [19,111,144,145]. LCA has often been applied in calculating environmental effects and checking sustainability [104,136,146], which is generally carried out by following the ISO standards 14040:2006 and 14044:2006 [118,119].

The scope of this research work is the inventories collected from the database for the considered plants. LCA is performed using SimaPro software. The analysis methods selected are ReCiPe 2016, Impact 2002+,

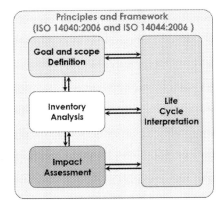

Figure 4.3 Stages of the LCA method [5].

Eco-points, and IPCC. These methods are widely used for life cycle environmental impact evaluation related to energy production technologies [120,147,148]. In this study, LCA is carried out in four fundamental steps:
1. goal and scope definition, where the aim is depicted and boundaries are identified;
2. LCI, where the energy-, material-, and emission-based input–output flows are assembled;
3. life cycle impact estimation, where effects are estimated for 14 impact indicators;
4. impact outcome interpretation, where effects are justified with the goal.

These key steps are shown in Fig. 4.3 and highlighted in the following subsections to depict the LCA method used in this research.

4.3.1 Goal and scope definition

In the first step of LCA, the main aim is the comparative evaluation of the environmental threats of the hydropower plants of alpine and nonalpine zones. It helps to recognize the best hydropower generation plants' locations in Europe from an environmental perspective. LCA is performed considering the cradle-to-grave LCA boundary for both locations. Therefore, the total inputs and outputs over the whole lifespan of the plants are considered in assessing the impacts, which are obtained from the database. Figs. 4.4 and 4.5 show the rates of energy and materials flow in various stages of the hydroelectricity generation in alpine and nonalpine areas, respectively. The functional unit is considered as 1 MJ of energy production

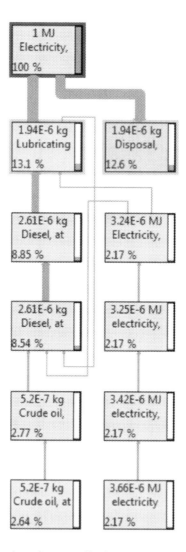

Figure 4.4 Materials flow sheet for 1 MJ of hydropower generation in an alpine region.

[100,113]. This functional unit of 1 MJ reveals the amount of electrical energy which is available for consumption at the distribution terminal.

4.3.2 Life cycle inventory

The second step of LCA is the development of an LCI counting all the input and output materials, energies, and emissions. For that purpose, an

Environmental impacts of hydropower plants

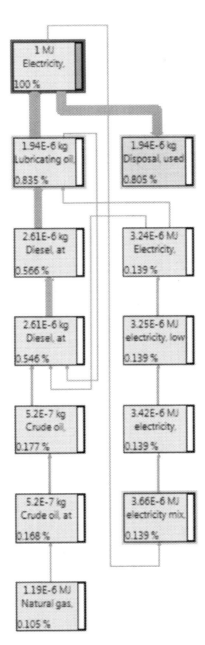

Figure 4.5 Materials flow sheet for 1 MJ of hydropower generation in a nonalpine region.

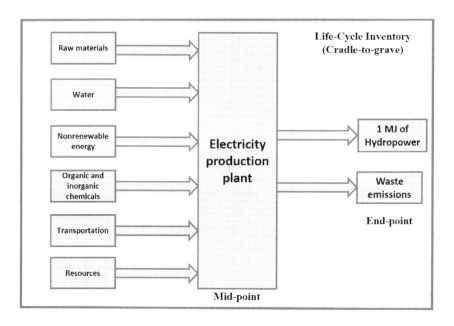

Figure 4.6 LCA system boundary used in this research.

LCA system boundary is considered in this research as depicted in Fig. 4.6. The inventory datasets of considered hydropower plants located in alpine and nonalpine regions were originally documented in [149], and were later compiled into the ecoinvent version 2 database. These datasets are now available in the Australian LCA database AusLCI. The datasets contain estimated values taking the average of more than 50 reservoir hydropower plants rated from 0.5 to 1200 MW in alpine and nonalpine areas of Europe. The weighted value of hydropower production is utilized to calculate the average material requirements. Raw materials, resources, and organic and inorganic chemicals are used to build the elements of hydropower plants. These plant elements along with construction materials (cement, gravel, concrete, and steel) are used to build the plants. The LCI includes all equipment for plant infrastructure construction, operation, and waste emissions. Fig. 4.6 shows the LCA system boundary considered in this LCA process.

The total inputs of a plant are of various types: raw materials, water, nonrenewable energy, organic and inorganic chemicals, transportation, and other available resources at the specific plant location. However, the output is 1 MJ of electricity production and waste emissions. Tables 4.4 and 4.5 show inputs from nature and the technosphere and outputs into the air

Table 4.4 LCI for LCA of the considered hydropower plants located in alpine regions.

Electricity, hydropower, at reservoir power plant, alpine region/RER U/AusSD U	1	MJ
Inputs from nature	Amount	Unit
Transformation, from unknown	2.3×10^{-5}	m^2
Transformation, to water bodies, artificial	2.28×10^{-5}	m^2
Transformation, to industrial area, built up	2.3×10^{-7}	m^2
Occupation, water bodies, artificial	3.5×10^{-3}	m^2
Volume occupied, reservoir	0.15	m^3
Water, turbine use, unspecified natural origin	0.81	m^3
Energy, potential (in hydropower reservoir), converted	3.79	MJ
Inputs from technosphere		
Lubricating oil, at plant/RER U/AusSD U	7×10^{-6}	kg
Emissions to air		
Dinitrogen monoxide	7.7×10^{-8}	kg
Methane, biogenic	1.4×10^{-5}	kg
Outputs to technosphere		
Disposal, used mineral oil, 10% water, to hazardous waste incineration/CH U/AusSD U	7×10^{-6}	kg

and the technosphere by the considered plants, respectively, collected from the LCI datasets. They show that the input amounts are almost identical for both plants. They also reveal that the amount of mineral oil disposal is identical for both systems, but nonalpine region-based hydropower plants produce more biogenic methane than alpine area-based installations.

4.3.3 Life cycle impact estimation

The third step of LCA is life cycle impact assessment, done following ISO 14044:2006, where emissions and input parameters are arranged into their respective impact indicators and converted into the same unit for comparative assessment. To calculate impacts, SimaPro software version 8.5 [138,150] and the ecoinvent database were used for the considered plants due to their global acceptance [121,122]. The ReCiPe 2016 method is used to assess the mid-point environmental impacts under categories such as global warming, ozone formation, ecotoxicity, water consumption, etc. This method is the updated version of ReCiPe 2008 by Huijbregts et al., which considered 17 mid-point effect indicators in characterizing the overall impacts of a system [151]. It characterizes the environmental effects of a

Table 4.5 LCI for LCA of the considered hydropower plants located in nonalpine regions.

Electricity, hydropower, at reservoir power plant, alpine region/RER U/AusSD U	1	MJ
Inputs from nature	Amount	Unit
Transformation, from unknown	2.3×10^{-4}	m^2
Transformation, to water bodies, artificial	2.28×10^{-4}	m^2
Transformation, to industrial area, built up	2.3×10^{-6}	m^2
Occupation, water bodies, artificial	3.5×10^{-2}	m^2
Volume occupied, reservoir	0.15	m^3
Water, turbine use, unspecified natural origin	8.1	m^3
Energy, potential (in hydropower reservoir), converted	3.79	MJ
Inputs from technosphere		
Lubricating oil, at plant/RER U/AusSD U	7×10^{-6}	kg
Emissions to air		
Methane, biogenic	2.86×10^{-4}	kg
Outputs to technosphere		
Disposal, used mineral oil, 10% water, to hazardous waste incineration/CH U/AusSD U	7×10^{-6}	kg

product through transforming LCIs into a number of impact categories by a unique state-of-the-art approach.

The Impact 2002+ method is used to measure the end-point effects under four categories: resources, climate change, ecosystem quality, and human health [123]. Human damage indicators are estimated considering carcinogenic and noncarcinogenic emissions. Furthermore, impacts related to human toxicity and ecotoxicity are obtained based on average responses instead of assumptions.

The IPCC method is used to evaluate the GHG emissions over a 100-year time span. This method considers emissions like carbon dioxide, methane, nitrous oxide, etc. [100,128]. This approach offers three benefits in assessing GHG emissions: (i) it ensures optimal use of the data source in a comprehensive manner, (ii) it establishes transparency in the assessment, and (iii) it provides insights for policymakers into climate solutions [152]. It does not consider carbon monoxide emissions and radiative releases in the stratosphere. Fig. 4.7 shows an overview of the LCA methods (ReCiPe 2016, Impact 2002+, and IPCC) used in this research.

The fossil fuel-based energy consumption rates by the plants have been evaluated by the CED approach. The CED method considers all types

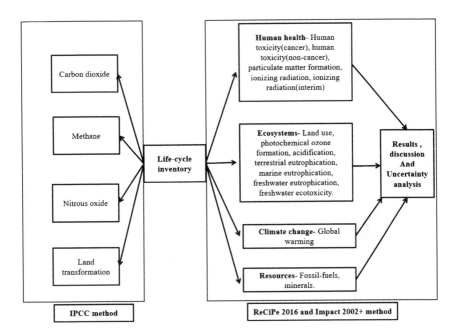

Figure 4.7 LCA methods used in this research.

of energy usage such as nuclear, biomass, renewable, and fossil fuel (oil, coal, and gas) during the overall life cycle of the plants and provides a breakdown of energy consumption [125]. It is pivotal to get information on the fossil fuel consumption by the plant to replace this by renewable energy to improve the environmental profile. Hydropower plants usually consume a small amount of fossil fuel during production and maintenance. However, they consume a significant amount of carbon-based fuel during raw material extraction for the manufacture of the plant elements [11]. Finally, uncertainty analysis has been performed by the Eco-indicator 99 method [153] to check the probability distribution of the considered cases for both plants.

4.3.4 Impact outcome interpretation

The final step of LCA is the life cycle impact interpretation, where effect outcomes are analyzed and compared with the main LCA goal of this research. At this stage, we identify the factors that are responsible for environmental hazards for each of the plants. Uncertainty analysis has been undertaken to interpret the LCA outcome in a broader perspective.

4.4. Results

4.4.1 Environmental profiles of hydropower plants

The environmental performances of hydropower plants in alpine and nonalpine regions of Europe are measured by the ReCiPe 2016 and Impact 2002+ methods [120]. The ReCiPe 2016 method is used to measure and compare the mid-point effects of the plants considering raw material extraction to electricity production, while the Impact 2002+ method is used to estimate and compare the end-point impacts of the plants considering raw material extraction to end-of-life waste disposal.

4.4.1.1 Mid-point impact assessment outcome

Fig. 4.8 shows the effects of the considered plants on global warming as determined by the ReCiPe 2016 approach. The global warming-related emissions mostly affect human health. The hydropower plants of nonalpine zones (weighted value of 4.16×10^{-9} DALY) have a stronger effect for the global warming human health indicator than those of alpine zones (weighted value of 1.78×10^{-9} DALY). This is because of the higher rate of methane biogenic emissions from nonalpine plants (2.51×10^{-9} DALY) than from alpine plants (1.23×10^{-10} DALY). However, the dinitrogen monooxide release is greater from hydropower plants in alpine regions (5.93×10^{-12} DALY) than from those in nonalpine regions (3.05×10^{-15} DALY). The impacts on terrestrial and freshwater ecosystems are also higher for hydropower plants in nonalpine areas (7.56×10^{-12} species.yr) than for those in alpine areas (2.07×10^{-16} species.yr) due to the methane biogenic emission rates.

The ozone formation-based impacts under the human health and terrestrial ecosystems categories are depicted in Fig. 4.9 for both plant locations. Both of the plants showed the same effects for ozone formation, with a weighted value of 4.87×10^{-12} species.yr for human health and 7.01×10^{-13} species.yr for terrestrial ecosystems.

The marine, freshwater, and terrestrial ecotoxicity-related impacts are highlighted in Fig. 4.10, which are also obtained by the ReCiPe 2016 method. For all three ecotoxicity-based impact categories, both systems revealed identical effects, with a weighted value of 5.14×10^{-16} species.yr for marine, a weighted value of 1.46×10^{-15} species.yr for freshwater, and a weighted value of 5.59×10^{-14} species.yr for terrestrial ecosystems.

Fig. 4.11 shows the water consumption-related impacts of the plants, obtained using the ReCiPe 2016 approach. The water consumption-

Environmental impacts of hydropower plants

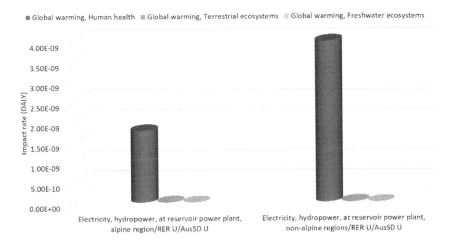

Figure 4.8 Global warming-based impact outcome comparison.

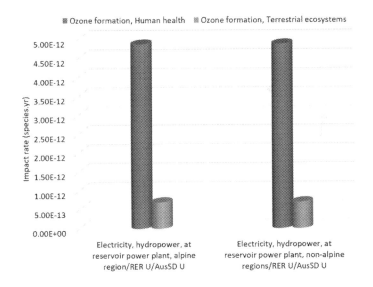

Figure 4.9 Ozone formation-based impact outcome comparison.

based impacts mostly affect human health (a weighted rate of 5.09×10^{-7} species.yr) rather than ecosystems (aquatic and terrestrial). Under this category, hydropower plants of alpine regions offer superior environmental profiles to nonalpine ones for all three impact indicators. The outcome suggests that water consumption (human health) is higher for hydropower

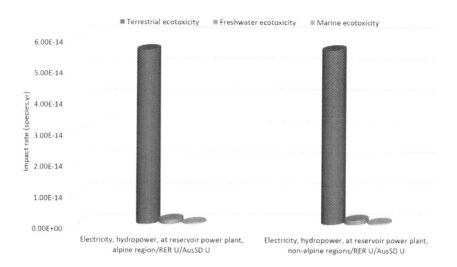

Figure 4.10 Ecotoxicity-based impact outcome comparison.

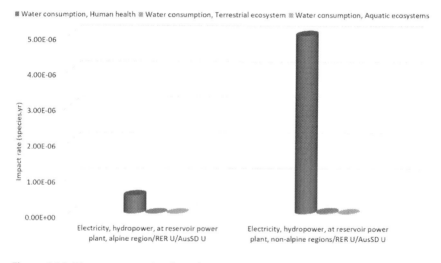

Figure 4.11 Water consumption-based impact outcome comparison.

plants in nonalpine regions due to the water consumed for turbine use, that is, 2.25 m³, whereas it is 0.225 m³ for hydropower plants in alpine regions.

The effects of other impact categories determined using the ReCiPe 2016 method, like land use, human carcinogenic and noncarcinogenic toxicity, fine particulate matter formation, ionizing radiation, stratospheric ozone depletion, freshwater eutrophication, and terrestrial acidification, are

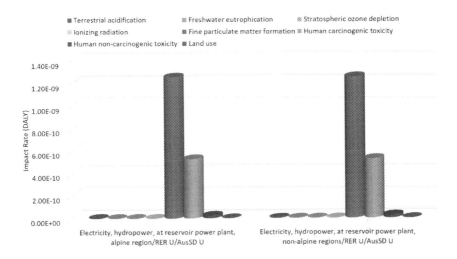

Figure 4.12 Effect outcome comparison for other impact indicators.

depicted in Fig. 4.12. The impacts are almost identical for plants in both locations. The highest impact category is the formation of fine particulate matter (weighted rate of 1.26×10^{-9} DALY), and the second greatest effect is observed for the human carcinogenic toxicity indicator (weighted rate of 5.28×10^{-10} DALY). The findings show that the formation of fine particulate matter resulted from sulfur dioxide emission (2.22×10^{-9} DALY for both locations).

4.4.1.2 End-point impact assessment outcome

The end-point environmental impacts of the considered plants are assessed by the Impact 2002+ method and the outcomes are obtained for 13 different impact indicators: carcinogens, noncarcinogens, respiratory inorganics, ionizing radiation, ozone layer depletion, respiratory organics, aquatic ecotoxicity, terrestrial ecotoxicity, terrestrial acid, land occupation, aquatic acidification, aquatic eutrophication, and global warming. All indicators effects except for global warming. Hydropower plants of alpine regions contribute less to global warming (8%) than nonalpine ones (100%).

All 13 categories assessed by the Impact 2002+ approach are categorized into four major types of effects: resources, climate change, ecosystem quality, and human health. Fig. 4.13 reveals that the plants of nonalpine areas (100%) contribute more to climate change than the plants of alpine areas (8%), while the impacts for other three major indicators were iden-

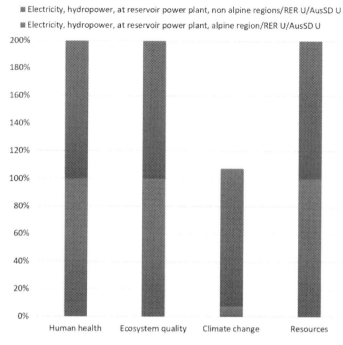

Figure 4.13 End-point damage assessment of the plants using the Impact 2002+ approach.

tical for both locations. This is because of the rates of methane biogenic and CO_2 emission into the air, which are 3.85×10^{-4} kg CO_2-eq. and 6.26×10^{-6} kg CO_2-eq., respectively, for the hydropower plants of non-alpine regions, whereas the methane biogenic emission is 1.89×10^{-5} kg CO_2-eq. and CO_2 release is 6.28×10^{-6} kg CO_2-eq. from the hydropower plants of alpine areas. For the resources, ecosystem quality, and human health impact categories, the major inventory outputs are from coal mining (8.15×10^{-12} MJ), nitrogen oxide (2.27×10^{-13} DALY), and sulfur dioxide (4.19×10^{-13} DALY), respectively.

4.4.2 Metal- and gas-based emission evaluation

The metal- and gas-based emissions are evaluated using the Eco-points 97 (CH) V2.07 method for both locations of hydropower plants. The Eco-points 97 method is a systematic way for evaluating the release of metal and hazardous gas by any product [115]. This method assesses 30 different indicators such as carbon dioxide (CO_2), nitrous oxide (NO_x), sulfur oxide (SO_x), ammonia (NH_3), nitrogen (N), copper (Cu), cadmium (Cd), nickel (Ni), mercury (Hg), zinc (Zn), lead (Pb), phosphorus (P), and chromium (Cr). The highest rate is set to 100%. Overall, the outcomes obtained by the Eco-points 97 method show that lower amounts of carbon dioxide gas (1.66×10^{-2} weighted points) are emitted from hydropower plants of alpine areas as compared to nonalpine ones (2.82×10^{-1} weighted points), whereas for other indicators the impacts are the same. A smaller amount of CO_2 is released from hydropower plants of alpine zones because of the standards in manufacturing the plants provided by The Swiss Federal Institute of Aquatic Science and Technology (EAWAG, the German acronym for *Eidgenossische Anstalt fur Wasserversorgung, Abwasserreinigung und Gewasserschutz*) [154] and the regular removal of sediments through flushing [132].

4.4.3 Greenhouse gas emission estimation

GHG release by the selected hydropower plants as determined using the IPCC approach [152] is depicted in Fig. 4.14. The GHG emissions are estimated for a 100-year time period. The hydropower plants of alpine and nonalpine areas release identical GHG emissions for the categories carbon dioxide and land transformation. A greater amount of methane (actual value of 1.7×10^{-3} kg CO_2-eq./MJ) is emitted by nonalpine plants than by alpine ones (8.18×10^{-5} kg CO_2-eq./MJ) due to the long transportation of raw materials and manufactured parts to build the overall system, whereas higher rates of nitrous oxide are emitted to the environment by alpine plants (6.64×10^{-6} kg CO_2-eq./MJ) than by nonalpine ones (7.7×10^{-9} kg CO_2-eq./MJ) due to the combustion of more fossil fuels during the manufacturing of parts and the combustion of more solid waste during end-of-life waste management [155]. These research findings will guide governments and energy investors in Europe to take well-informed decisions in installing hydropower plants in alpine regions.

4.4.4 Comparative life cycle inputs and outputs of the plants

The overall input–output comparison is performed by the Raw Material Flows (RMF) method. Fig. 4.15 shows the inputs from nature and the outputs into air, water, and soil and solid waste as determined using the RMF approach. All inputs and outputs are identical except for the outputs into air; alpine region-based plants emit less into the air than nonalpine-based plants. The emission rates into the air are higher for hydropower plants

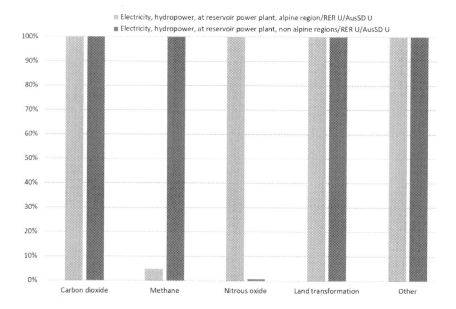

Figure 4.14 GHG emissions as determined by the IPCC approach.

in nonalpine regions due to higher emission rates of methane biogenic (7.94×10^{-5} kg for nonalpine plants and 3.89×10^{-6} kg for alpine ones), but for dinitrogen monooxide, hydropower plants of alpine regions have higher emission rates (2.14×10^{-8} kg for alpine plants and 1.1×10^{-11} kg for nonalpine ones). The weighted normalized values of the total inputs and outputs of alpine plants are about 3.98×10^{-6} and 1.19×10^{-5}, respectively. The weighted normalized values of the overall inputs and outputs of nonalpine plants are approximately 3.98×10^{-6} and 8.75×10^{-5}, respectively.

4.4.5 Energy consumption comparison

The fossil fuel-based energy consumptions of alpine and nonalpine hydropower plants are estimated using the CED approach [125]. The obtained outcomes are shown in Table 4.6. Among various kinds of energy, maximum consumption occurred from renewable sources with a normalized rate of 3.79 and 1.05 for alpine and nonalpine areas, respectively. This is because water is the main source of power used during operation of the plant. The use of nuclear and biomass power is very low for both locations. The hydropower plants of the alpine regions consume five times as much as the nonalpine hydropower plants. However, the coal-based energy usage by

Environmental impacts of hydropower plants

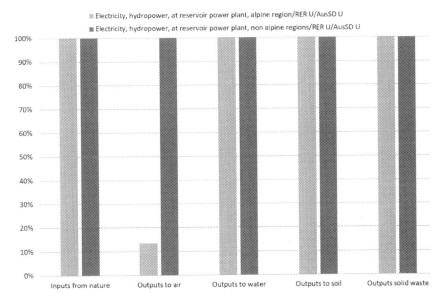

Figure 4.15 Comparative life cycle inputs and outputs of hydropower plants of alpine and nonalpine regions as determined by the RMF method.

Table 4.6 Life cycle energy consumption by the considered hydropower plants, as determined by the CED method.

Label	Electricity, hydropower, at reservoir power plant, alpine region/RER U/AusSD U [MJ]	Electricity, hydropower, at reservoir power plant, nonalpine regions/RER U/AusSD U [MJ]
Renewables	3.79	1.05
Fossil fuels – oil	5×10^{-4}	1×10^{-4}
Fossil fuels – gas	3.78×10^{-5}	1.05E-5
Fossil fuels – coal	3.03×10^{-5}	8.43×10^{-6}
Biomass	1.81×10^{-8}	5.03×10^{-9}
Nuclear	9.92×10^{-9}	2.76×10^{-9}

hydropower plants of the alpine zones is about one-third that of nonalpine ones. Overall, the outcome indicates that carbon-based power consumption over the lifetime of the considered plants creates a negative impact on the environment. Therefore, future research should be carried out to reduce the use of fossil fuel-based energy by replacing it with renewable energy.

Table 4.7 Key impact comparison with previous studies.

Country	Plant capacity (MW)	Major impact category	GHG release (kg CO_2-eq./kWh)	Dominant material	Ref.
Europe (alpine)	0.5–1200	GWP	1.07×10^{-4}	Construction	This study
Europe (nonalpine)	0.5–1200	GWP	1.41×10^{-3}	Transmission line	This study
Brazil	30.3	GWP	5.46	Transport	[87]
Canada	10	GWP	15.2	Construction	[56]
China	3.2	GWP	7.6	Reservoir	[75]
India	3	GWP	74.88	Turbine	[58]
Indonesia	8	ETP	1.2	Pipeline	[86]
Japan	10	GWP	11.3	Construction	[135]
Thailand	3	GWP	52.7	Transmission line	[74]

4.5. Discussion
4.5.1 Impact outcome comparison with previous studies

The construction phase is responsible for most impacts by hydropower plants in alpine areas, whereas the transmission line is responsible for most impacts by hydropower plants in nonalpine areas (Table 4.7). According to Siddiqui et al., the construction and decommissioning phases are the main contributors to environmental impacts [56]. If biomass decay is considered, it may be a significant contributor to global warming. The obtained human toxicity and photochemical ozone creation values are 0.0047 g dichlorobenzene-eq./kWh and 0.000768 g ethane-eq./kWh, respectively, whereas the global warming potential value is 15.2 g CO_2-eq./kWh. Hanafi et al. showed that the highest impact of a hydropower plant in Indonesia is on marine aquatic ecotoxicity, freshwater ecotoxicity, and abiotic depletion [86]. The main sources of these impacts are rapid pipelines (59%) and building construction (20.5%). The obtained GHG emission rate is 1.2 kg CO_2-eq./MWh.

However, according to Li et al. the GHG emission rate of a hydropower plant in China is 7.6 ± 1.09 g CO_2-eq./kWh [15]. The reservoir and dam are the most sensitive factors for GHG emissions. The potential release of methane from sediment in the phase of dam decommission plays a crucial part in its overall impact. Pascale et al., who conducted a cradle-to-grave

LCA analysis of a hydropower plant in Thailand, found that the global warming potential and the photochemical ozone creation rate are 52.7 g CO_2-eq./kW and 0.03 g ethane-eq./kW, respectively [74], whereas the amount of GHG released by hydropower plants in the nonalpine areas is 1.41×10^{-3} kg CO_2-eq./kWh. The transmission line is the dominant component in almost all life cycle impact assessment categories with the exception of abiotic depletion. The distribution network and penstock are also responsible for a large percentage of the environmental impacts. Geller et al. showed that the global warming potential, human toxicity, and freshwater ecotoxicity of a hydropower plant in Brazil are 5.46 kg CO_2-eq./MW, 7.28 kg 1,4-dichlorobenzene-eq./MW, and 2.45 kg 1,4-dichlorobenzene-eq./MW, respectively [87]. Transport is the most crucial emitter since most of the equipment and materials for plant construction were brought from distant locations, through various means of transportation.

The key finding of Pang et al. in China is that the main impact of a hydropower plant occurs in the construction phase [75]. The small hydropower plants of Thailand and Japan work similarly to those of China, but worse than plants in Switzerland. According to this research, for the considered case the global warming potential value is 28.4 kg CO_2-eq./MWh, the abiotic depletion value is 91.6 g antimony-eq./MWh, the freshwater ecotoxicity value is 11.1 kg dichlorobenzene-eq./MWh, and the photochemical ozone creation rate is 9.3 g C_2H_4-eq./MWh. About 96.1% of the global warming potential is incurred at the construction stage. Hondo et al. from Japan discovered that construction of hydropower plants is the dominant phase for the global warming potential, which constitutes 82.8% of total GHG emission. The amount of global warming potential by the considered hydropower plant in Japan is 11.3 kg CO_2-eq./kWh [135], whereas the GHG emission rate by the hydropower plant of the alpine area in Europe is 1.07×10^{-4} kg CO_2-eq./kWh. The difference in impact outcome between the hydropower plants of different countries is due to differences in available resources, raw materials, transportation distances of plant elements, and combustion rates of fossil fuels during manufacture of parts and end-of-life waste management. Overall, for a typical hydropower plant, the construction phase is responsible for the impacts in all categories, the operation phase is responsible for abiotic depletion and freshwater ecotoxicity, and transportation is responsible for acidification.

4.5.2 Impact comparison with other power plants

According to the end-point indicator-based results obtained by the Impact 2002+ method, electricity generated by hydropower plants in both alpine and nonalpine regions shows greater environmental sustainability than electricity generated by other sources like lignite coal, natural gas, biomass, wind, and photovoltaic (PV) panels. From the eight major impact categories, electricity generated by lignite coal has the most detrimental effect overall for all impact categories except noncarcinogens and ozone layer depletion (Fig. 4.16 and Table 4.8); for the noncarcinogens and ozone layer depletion categories, electricity generated by natural gas has the most adverse effect. Electricity generation from wind results in more ionizing radiation than other sources. Electricity generation from biomass has lower impacts like hydropower for most of the impact indicators. Alpine and nonalpine hydropower plants have very similar impacts on the environment in all categories except global warming. Nonalpine hydropower plants have stronger effects in terms of global warming. A similar scenario is observed from Table 4.9, which shows a comparison of power plants based on damage categories. Coal-based power plants affect human health, ecosystem quality, climate change, and resources more than other power plants. Overall, hydropower plants in alpine regions are better in terms of climate change than other plants.

4.5.3 Uncertainty analysis

The uncertainty distributions of the two estimations for the considered hydropower plants as determined by the Eco-indicator 99 method are shown in Figs. 4.17 and 4.18. The bars with maximum height show the greatest probabilities. The estimated life cycle single-score points of alpine and nonalpine hydropower plants are $1.18 \times 10^{-4} \pm 3.34 \times 10^{-5}$ (mean $\pm$ standard deviation) and $1.33 \times 10^{-4} \pm 1.43 \times 10^{-4}$, respectively. The smaller standard deviation in the single-score points of alpine hydropower plants indicates lower impact probabilities than for nonalpine plants. Overall, the small bars for about 70% of the total numbers indicate that hydropower plants of both regions are environmentally superior to other plants.

4.6. Limitations and future improvements

The main limitations and future improvements of this research are outlined as follows:

Environmental impacts of hydropower plants

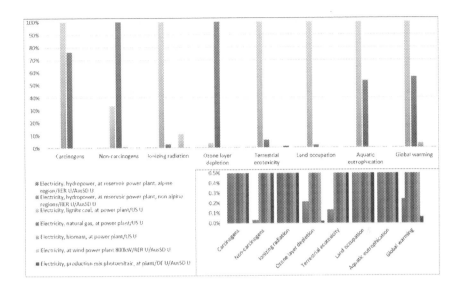

Figure 4.16 Environmental impacts of various power plants.

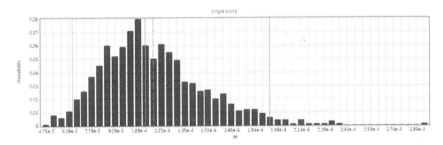

Figure 4.17 Probability distribution for the single-score impact category of hydropower plants of alpine zones.

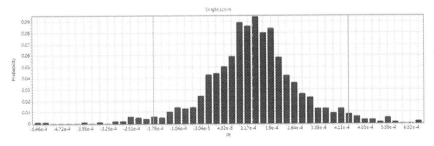

Figure 4.18 Probability distribution for the single-score impact category of hydropower plants of nonalpine zones.

Table 4.8 Key impacts of various plants.

Impact category	Electricity, hydropower, at reservoir power plant, alpine	Electricity, hydropower, at reservoir power plant, nonalpine	Electricity, lignite coal, at power plant	Electricity, natural gas, at power plant	Electricity, biomass, at power plant	Electricity, at wind power plant	Electricity, production mix photovoltaic, at plant	Unit
Carcinogens	2.18×10^{-8}	2.18×10^{-8}	3.12×10^{-4}	2.38×10^{-4}	8.33×10^{-7}	7.35×10^{-7}	4.01×10^{-8}	kg C_2H_3Cl-eq.
Noncarcinogens	5.89×10^{-8}	5.89×10^{-8}	5.89×10^{-3}	1.76×10^{-2}	1.64×10^{-4}	1.37×10^{-6}	1.05×10^{-6}	kg C_2H_3Cl-eq.
Ionizing radiation	2.08×10^{-8}	2.08×10^{-8}	4.47×10^{-6}	1.28×10^{-7}	8.56×10^{-9}	4.98×10^{-7}	3.08×10^{-9}	Bq C-14-eq.
Ozone layer depletion	1.17×10^{-12}	1.17×10^{-12}	2.12×10^{-9}	5.41×10^{-8}	7.02×10^{-11}	2.80×10^{-11}	1.20×10^{-13}	kg CFC-11-eq.
Terrestrial ecotoxicity	5.37×10^{-6}	5.37×10^{-6}	4.02×10^{-1}	2.47×10^{-2}	5.01×10^{-4}	1.81×10^{-4}	4.70×10^{-3}	kg
Land occupation	3.01×10^{-8}	3.01×10^{-8}	5.62×10^{-4}	1.23×10^{-5}	1.07×10^{-6}	7.04×10^{-7}	3.91×10^{-7}	m^2
Aquatic eutrophication	9.466×10^{-10}	9.46×10^{-10}	1.28×10^{-5}	6.81×10^{-6}	6.31×10^{-8}	2.22×10^{-8}	2.29×10^{-8}	kg
Global warming	2.97×10^{-5}	3.92×10^{-4}	1.21	6.84×10^{-1}	4.45×10^{-2}	1.81×10^{-4}	7.59×10^{-6}	kg CO_2-eq.

Table 4.9 Key damage comparison with various plants.

Damage category	Electricity, hydropower, at reservoir power plant, alpine region/RER U/AusSD U	Electricity, hydropower, at reservoir power plant, nonalpine regions/RER U/AusSD U	Electricity, lignite coal, at power plant/US U	Electricity, natural gas, at power plant/US U	Electricity, biomass, at power plant/US U	Electricity, at wind power plant 800 kW/RER U/AusSD U	Electricity, production mix photovoltaic, at plant/DE U/AusSD U	Unit
Human health	1.67×10^{-12}	1.67×10^{-12}	9.12×10^{-7}	4.20×10^{-7}	1.31×10^{-7}	4.11×10^{-11}	6.36×10^{-12}	DALY
Ecosystem quality	1.22×10^{-7}	1.22×10^{-7}	3.73×10^{-2}	1.48×10^{-2}	4.79×10^{-3}	3.37×10^{-6}	4.02×10^{-5}	m²
Climate change	2.97×10^{-5}	3.92×10^{-4}	1.21	6.84×10^{-1}	4.45×10^{-2}	1.81×10^{-4}	7.59×10^{-6}	kg CO_2-eq.
Resources	8.15×10^{-12}	8.15×10^{-12}	20.52	12.63	3.42×10^{-2}	1.95×10^{-10}	8.19×10^{-12}	MJ

- Pumped storage plants are not separated during the estimation of reservoir plants average rates at the used data source.
- Environmental impacts of individual components of the hydropower plants have not been assessed separately in this work.
- The environmentally impactful elements can be replaced by sustainable alternatives, which must be assessed by further sensitivity analysis.
- Life cycle cost analysis and techno-economic analysis, which can be of great value in the future, have not been conducted in this study due to a lack of sufficient data.

The future direction of this research work would be to solve the above-mentioned issues and make hydropower production greener in the near future.

4.7. Conclusion

In conclusion, a comparative assessment of the environmental impacts of an alpine and a nonalpine hydropower plant is highlighted. To confirm the efficacy of this LCA analysis, (i) a new comprehensive LCI for both types of plants is developed, (ii) impacts are assessed by several methods, and (iii) an uncertainty analysis of both frameworks is carried out to judge their effectiveness. This research work is unique as it quantifies the ecological effects by several systematic methods for the first time and considers both mid-point and end-point approaches in assessing the effects from both installations, utilizing the Impact 2002+ method. The novel findings of this research are that hydropower plants in alpine regions are environment-friendly and have only 8% carbon dioxide emission with a smaller effect on climate change, while hydropower plants in nonalpine regions have high carbon dioxide emissions with a greater effect on climate change. The main result obtained using the IPCC, ReciPe 2016, and Impact 2002+ methods is that the effects of hydropower plants in nonalpine regions on global warming (3.92×10^{-4} kg CO_2-eq./MJ) are stronger than those of plants in alpine regions (2.97×10^{-5} kg CO_2-eq./MJ). Moreover, alpine plants cause one-tenth the methane release of nonalpine plants. However, both alpine and nonalpine hydropower plants are more environment-friendly electricity generation sources than lignite coal, natural gas, biomass, wind, and PV panels. Overall, to stimulate clean hydropower generation, future research should be focused on finding the most hazardous materials in plant installations and replacing them with analogous ecologically superior alternatives.

CHAPTER FIVE

Environmental impact assessment of renewable power plants in the US

Renewable electricity generation technologies provide a more sustainable solution than nonrenewables. However, as complete systems, renewable energy generation systems have impacts on humankind, resources, and ecosystems. This chapter will address the environmental effects of different types of renewable plants through life cycle assessment (LCA). A comparative study is performed among solar-photovoltaic (PV), biomass, and pumped storage hydropower plants in the US. Life cycle impact analysis has been carried out by the Eco-indicator 99, Tool for the Reduction and Assessment of Chemical and other Environmental Impacts (TRACI), Raw Material Flows (RMF), cumulative energy demand (CED), Eco-points 97, and Intergovernmental Panel on Climate Change (IPCC) methods, using SimaPro software. The impacts are considered based on 10 midpoint impact categories and 3 end-point indicators. The results show that pumped storage hydropower plants have the highest environmental impacts in the categories of global warming, ozone depletion, acidification, eutrophication, carcinogenics, and ecotoxicity, while biomass plants have higher impacts in the categories of smog and fossil fuel depletion. Moreover, pumped storage hydropower plants have the highest impact on human health (7.74E−05/kWh) and ecosystems (3.35E−06/kWh), whereas biomass plants have higher effects on resources (4.79E−07/kWh). Overall, solar-PV plants are found to be much more environment-friendly than other renewable electricity generation systems. These findings will guide investors in installing sustainable and clean power plants.

5.1. Introduction

With the growth of the world population and the economy, the demand for energy has increased enormously. To fulfill this rising demand, increasing amounts of fossil fuel are burned at power plants, which produce hazardous chemicals and pollute our environment [100,130]. Globally about 900 tons of CO_2 are released into the air every second by fossil fuel-based power plants [30]. People are getting more conscious of the adverse ecological impact of conventional power generation units. To reduce pollution, in the last few decades renewable energy technologies (RETs) have become popular [28,29,97,104,156]. Moreover, RETs produce electricity with enhanced cost-effectiveness, increased efficiency, and superior environmental profiles [38,96,157]. Most of the research on RETs has focused on productivity enhancement and cost–benefit analyses [33,158]. However, renewable power plants also cause emissions [15,53,94,107,159], mostly during the extraction of raw materials, production of elements from the extracted raw materials, transportation of the materials to the plant location, and the end-of-life waste management period [24,48,109,148,160]. These emissions from RETs are lower than those of conventional power generation units, but not negligible [58,161–163].

The shift from conventional to sustainable energy production would be very beneficial for our environment. Energy optimization-based models are not suitable to make this shift. An LCA-based study carried out by Siddiqui et al. highlighted a comparison of the environmental impacts of nuclear, wind, and hydropower plants in Ontario, raising problems like variations in life cycle inventories (LCIs) and modeling approaches, and therefore the lack of a complete depiction of the total effects [56]. Furthermore, they have not considered solar-PV and biomass power plants, and their research is only applicable to Ontario, as global data have not been used. Another comparative LCA-based environmental effect assessment of solar-PV and wind energy systems was done by Nugent et al., who measured and compared the life cycle greenhouse gas (GHG) emissions; other impacts have not been considered [53]. Some previous studies only analyzed the effects of one element of the overall plant, like polymer solar cells [22,23,164] or PV panels [25], and focused on reducing the amount of hazardous raw material used in order to minimize the impact. A comprehensive LCI was not included in either of the abovementioned LCA analyses, so they lack LCA modeling accuracy and reliability. Separate LCA research was conducted on PV systems [102,103,131], wind turbines [165], hydropower plants [35,134], and biomass power plants [47,51] by different research groups, but none quantified the amount of fossil fuel-based energy consumption by the considered plants during construction, usage, and end-of-life management. Therefore, research must be carried out in this field to find the exact emission rates during each production step of the plant elements and to discover ways to reduce those emissions. Country-wise LCA of different renewable production technologies was carried out by various researchers who have not considered global databases like ecoinvent in life cycle input–output measurement [27,32,41,57,100,166,167]. Table 5.1 highlights the previous country-based LCA work for renewable technologies such as solar-PV, pumped storage hydropower, and biomass plants. The choice of a power plant in different nations depends on the availability of the required resources, their abundance, and reliability. These resources vary among countries based on their geographic position, latitude–longitude, and resources; this chapter thus compares renewable electricity generation systems in the US because the US has notable solar-PV, biomass, and pumped storage hydropower plants. For that reason, this research aims to highlight the effects of these plants on the environment through a dynamic LCA approach.

LCA is a useful tool to quantify the environmental impact of any product, unit, or system utilizing different methods considering the material flows, input resources, and output emissions [9,13,59,120,129,168–171]. The ecoinvent database is utilized to collect the material flows and gather all the life cycle input and output data for creating a new LCI and system boundary for all of the considered plants. LCA is carried out using SimaPro software by the Eco-indicator 99, RMF, TRACI, CED, Eco-points 97, and IPCC methods. In brief, the key contributions of this research work are as follows:

- Assessment and comparison of the mid-point and end-point indicator-based environmental impacts of three different RET-based power plants: a solar-PV power plant, a pumped storage hydropower plant, and a biomass power plant.
- Evaluation of hazardous metals and GHG emissions from the considered plants over their lifetime.
- Estimation of the amount of fossil fuel-based energy consumption during the manufacturing of plant elements, the installation and operation of the plants, and the waste management period.
- Uncertainty analysis for all considered cases.

This research is superior to previous LCA-based environmental effect assessments of renewable energy systems in several ways; for example, a comprehensive LCI is created using a reliable global database, the impacts for about 10 mid-point effect indicators including GHG emissions are assessed, and the fossil fuel-based energy consumption rate of each plant is taken into account. Considering all of these aspects, this is a pioneering

Table 5.1 Country-based overview of previous research on solar-PV, hydro, and biomass power plants.

Country	Topic	Key contributions	Research gaps	Ref.
Singapore	A comparative LCA of PV electricity generation in Singapore by multicrystalline silicon technologies	The environmental impact of three roof-integrated PV systems in Singapore in terms of GHG and energy payback time as the assessing indicators is demonstrated. Moreover, this research has found that PV modules with frameless double glass can enhance the environmental superiority.	In this research, only PV modules of silicon are considered for various case studies. The result might be different for other types of PV modules such as copper indium gallium selenide, cadmium tellurium, etc.	[48]
China	LCA of grid-connected PV power generation from crystalline silicon solar modules in China	The energy consumption and GHG emissions of a solar-PV system in China are estimated, where PV modules are built from crystalline silicon.	The energy consumption rates during the production of elements, transportation, construction, and manufacturing stages of the plant have not been assessed to analyze the overall fossil fuel usage by the plant.	[107]
Australia	Environmental impacts of solar-PV and solar-thermal systems with LCA	The environmental superiority of solar-PV over thermal has been shown by different LCA methods.	Sensitivity analysis has not been done for all of the plant elements of the solar-PV system.	[10]

continued on next page

Table 5.1 (continued)

Country	Topic	Key contributions	Research gaps	Ref.
Brazil	LCI for hydroelectric generation: a Brazilian case study	A new LCA inventory has been designed for hydropower-generating plants in Brazil.	The LCA estimation has not covered all the environmental impact indicators for the plant.	[57]
China	Carbon footprints of two large hydroprojects in China: LCA according to ISO/TS 14067	A new system boundary for LCA of two hydroelectricity production plants is designed considering installation, operation, maintenance, and decommissioning of the dam.	In this research only reservoir-based hydropower plants have been considered for overall impact assessment.	[15]
Ecuador	Analysis of GHG net reservoir emissions of hydropower plants in Ecuador	The GHG emissions of two reservoir-based hydropower plants are assessed considering all possible sources of emissions.	The social aspects have not been included in this LCA analysis.	[36]
France	LCA applied to electricity generation from renewable biomass	A comparative impact assessment between 2-MW and 10-MW biomass plants has been carried out.	An assumption is made in estimating the cogeneration performance of the plants which is more regulatory, and the impacts associated with nonregulated pollutants have not been considered in this research.	[51]

continued on next page

Table 5.1 (continued)

Country	Topic	Key contributions	Research gaps	Ref.
South Africa	LCA of fossil carbon dioxide emissions reduction scenarios in coal biomass-based electricity production	The land use and occupation impacts considering relevant processes of biomass combustion are estimated.	The impacts of land occupation and transformation have not been evaluated.	[85]
Indonesia	Techno-economic and GHG savings assessment of decentralized biomass gasification for electrifying rural areas in Indonesia	The impact of a biomass gasification system in Indonesia on global warming is evaluated using the LCA approach.	This LCA analysis is limited to a small scenario of a biomass plant, which must be carried out considering regions or countries using hybrid LCA for various biomass types.	[47]

work that highlights and compares the total environmental effects of solar-PV, biomass, and pumped storage hydropower plants located in the US.

In light of the above, the rest of the chapter is organized as follows. **Section 5.2** describes the generation- and consumption-based electricity matrix in the US. **Section 5.3** highlights the materials and methods to carry out a systematic LCA for assessing environmental profiles, analyzing GHG emission factors, and conducting uncertainty analysis. The overall LCA method is described. The results are presented and interpreted in **Section 5.4**, in four parts – the plants' environmental effects, evaluation of their fossil fuel-based energy consumption, estimation of the GHG release, and comparison with other types of plants located in the US. The uncertainty and sensitivity analyses are highlighted in **Sections 5.5** and **5.6**, respectively. **Section 5.7** discusses the overall impacts of the plants, compares these with existing studies, and highlights the limitations of this study. Finally, concluding comments on the research outcome are given in **Section 5.8**.

5.2. US electricity generation and consumption overview

Table 5.2 presents an overview of the energy production in the US based on different energy sources. Fossil fuels account for 2536 billion kWh (62.9% of the total energy). Of the fossil fuels, natural gas accounts for 1296 billion kWh (32.1% of the total) and coal accounts for 1206 billion kWh (29.9% of the total) [8]. The remaining fossil fuels (0.9%) include energy sources like petroleum, petroleum coke, and various gases. Nuclear energy accounts for 805 billion kWh (20% of the total). Renewables account for 687 billion kWh (17% of the total). Among the renewables, hydropower accounts for 300 billion kWh (7.4% of the total), wind energy accounts for 254 billion kWh (6.3% of the total), and biomass accounts for 63 billion kWh (1.6% of the total). Pumped storage hydropower production systems consume more energy to pump water to the storage than they generate, so the overall yearly generation is negative (−0.2% of the total).

Fig. 5.1 shows the energy consumption in the US based on different electricity sources between 1776 and 2017 [6]. The chart shows that before 1851, all power was from wood. In 1851, coal was introduced to produce power. Later the materials also included other fossil fuels and various types of renewables utilized for solar power generation.

5.3. Methodology

The main focus of this work is to assess the environmental threats of three different renewable power plants, namely solar-PV, biomass, and pumped storage hydropower, and compare their ecological consequences based on major impact categories. To fulfill this goal, a systematic LCA approach is utilized. LCA is considered to be the most powerful tool for assessing the environmental effects of any product, unit, process, or system of processes [111,113]. The application of LCA includes, but is not limited to, impact assessment, uncertainty analysis, and sustainability investigation [114–117,150]. This LCA approach is usually accomplished by following standards 14040:2006 and 14044:2006 of the International Standardization Organization (ISO) [118,119]. The LCA software SimaPro (version 8.5) is utilized to assess the environmental effects. LCA is performed by collecting the inputs and outputs at each manufacturing step of the different elements, and the ecological impacts are calculated utilizing four fundamental steps:

Table 5.2 Electricity production in the US based on different energy sources [8].

Energy source	Billion kWh	Share of total
Total – all sources	4034	
Natural gas	1296	32.1%
Coal	1206	29.9%
Petroleum (total)	21	0.5%
Petroleum liquids	12	0.3%
Petroleum coke	9	0.2%
Fossil fuels (total)	2536	62.9%
Nuclear	805	20.0%
Hydropower	300	7.4%
Wind	254	6.3%
Biomass (total)	63	1.6%
Wood	41	1.0%
Landfill gas	12	0.3%
Municipal solid waste (biogenic)	7	0.2%
Other biomass waste	3	0.1%
Solar-PV (total)	53	1.3%
PV	50	1.2%
Solar-thermal	3	0.1%
Geothermal	16	0.4%
Renewables (total)	687	17.0%
Pumped storage hydropower	−6	−0.2%
Other sources	13	0.3%

1. goal and scope definition for assessment, where the aim is defined and the LCA boundaries are set;
2. LCI analysis, where inputs and outputs at every step of the manufacturing processes are included;
3. life cycle environmental impact evaluation using the data compiled at the previous stage, where yield emissions and input assets are divided into their specific effect indicators and changed into the same units for comparative evaluation;
4. impact outcome interpretation and suggestions to achieve the goals of this study.

In the following subsections, the above LCA steps are described briefly to depict the LCA methodology which is followed in this research.

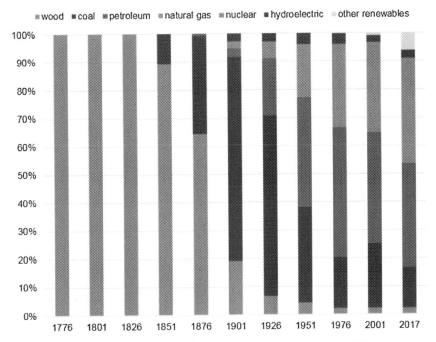

Figure 5.1 Electricity consumption overview in the US based on different energy sources [6].

5.3.1 Goal and scope definition

The primary goal of this LCA is to calculate the environmental threats of three different renewable energy generation systems, namely a solar-PV power plant, a biomass power plant, and a pumped storage hydropower plant. This LCA also compares their impact outcomes to identify the most environment-friendly renewable electricity generation plant. LCA is performed considering a cradle-to-gate perspective (from raw material extraction to disposal) for all of the considered systems [11,117,120]. Therefore, the overall LCA considers the complete life cycle of the plants, including stages such as raw material extraction and transportation to the production plant, input material manufacturing and transportation to the plant, plant installation and operation, and end-of-life recycling and disposal of elements. Figs. 5.2 to 5.4 sequentially illustrate the absolute and relative amounts of the energy and material flow in different steps of the solar-PV power plant, the pumped storage hydropower plant, and the biomass power plant to produce 1 kWh through their respective material flow sheet. The

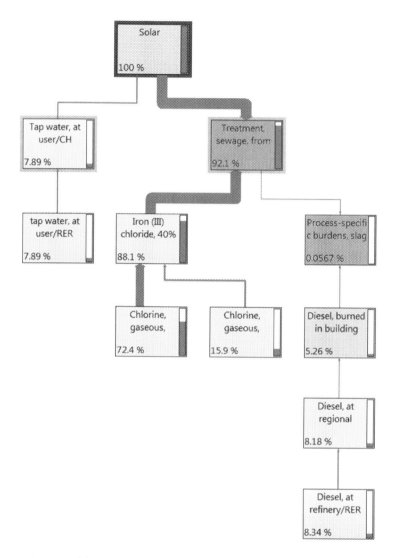

Figure 5.2 Material flow sheet for 1 kWh of solar-PV power generation.

functional unit is 1 kWh, which represents the amount of energy produced by the overall system for which the impacts are assessed [100,111,113].

5.3.2 Life cycle inventory

This second step of LCA is the collection of input raw materials, input energies, input resources, output emissions, end-of-life recycling quanti-

Environmental impact assessment of renewable power plants in the US

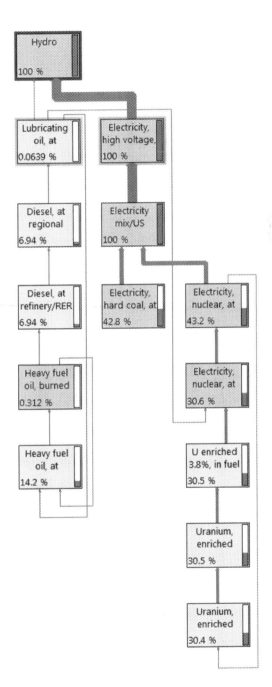

Figure 5.3 Material flow sheet for 1 kWh of pumped storage hydropower generation.

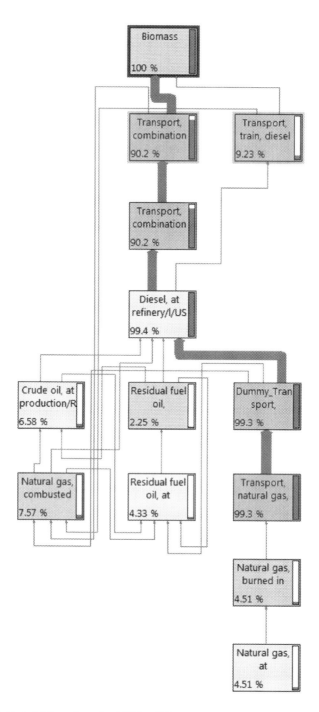

Figure 5.4 Material flow sheet for 1 kWh of biomass power generation.

Table 5.3 Data sources for the considered renewable power plants in the US.

Plant	Data source	Ref.
Solar-PV	Electricity, production mix photovoltaic, at plant/US U	[172]
Pumped storage hydropower	Electricity, hydropower, at pumped storage power plant/US U	[173]
Biomass	Electricity, biomass, at power plant/US U	[174]

ties, and disposal amounts per unit process considered in the LCA system boundary. To get all these rates, it is required to collect LCI data from any recognized database, industry, or literature review. The inventories for various renewable energy plants change among countries based on the location and available resources. Therefore, we collected data from the renowned ecoinvent database (version 3.5) [121,122,150] for solar-PV, biomass, and pumped storage hydropower plants located in the US. Table 5.3 shows the LCI data source for each of the studied plants in this research [172–174]. Basically, all the renewable power generation units follow almost the same steps throughout their lifetime such as extraction of raw materials from mines, transportation of the extracted raw materials to industry, manufacture of plant elements from the raw materials, transportation of raw materials to the plant, installation of the plant, production of electricity, and end-of-life waste management. For that reason, we considered a common system boundary for all three plants in our LCA analysis. Fig. 5.5 highlights the system boundary considered according to the impact assessment methods used here for analysis. The overall inputs of a plant are raw materials, water, nonrenewable energy, organic and inorganic chemicals, transportation, and other available resources at the specific plant location. On the other hand, the outputs are 1 kWh of energy production and waste emissions.

5.3.3 Life cycle impact assessment

LCA is carried out following the systematic steps of assessing impacts by different approaches and comparing the outcomes among the three considered plants. For forming the LCI, the ecoinvent database is used to gather the data for the plants [112,121,122]. The ecoinvent database contains international industrial LCI data for raw material extraction, transportation, energy usage, etc. SimaPro software version 8.5 is used to assess the impacts after creating the LCI and system boundary as it is widely used for LCA [138]. The overall input and output scenarios are assessed and compared

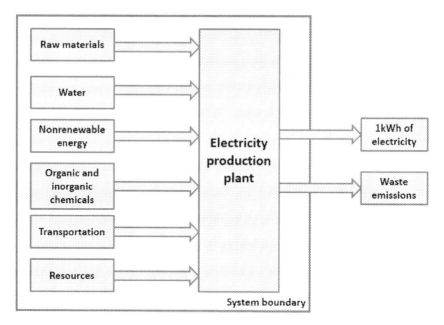

Figure 5.5 Common system boundary for all power generation processes used in this LCA analysis.

among all the considered renewable power plants by the RMF method [10]. This method tracks the total mass flow of raw materials and emission based on the summation of all elementary flows. In theory, if all inventories are mass-balanced, the sum of the inflows should equal the amount of the outflows, but this is rarely the case as combustion incorporating oxygen input is often not included, and moisture balances are also not always tracked.

The TRACI method is utilized to find out the cradle-to-gate environmental impacts for 10 impact indicators, as it elaborates the problem-oriented (mid-point) approach, invented by the US Environmental Protection Agency [175]. It reflects the state-of-the-art practices of LCA and depicts the potential effects in the context of the US. It provides potential ecological effects under impact categories such as fossil fuel depletion, ecotoxicity, respiratory effects, carcinogenics, noncarcinogenics, eutrophication, acidification, smog, global warming, and ozone depletion.

The end-point impacts of the plants are assessed by the Eco-indicator 99 method, which is updated from the 95 version [176]. This approach is a weighting method invented for sustainability modeling of products. It

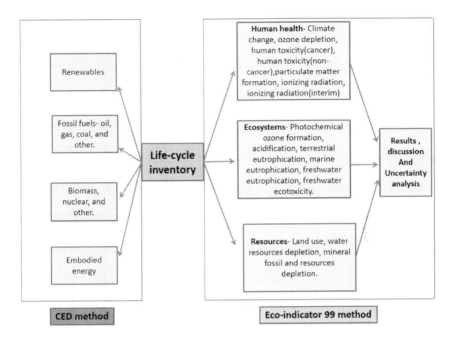

Figure 5.6 Life cycle impact assessment methods used in this research.

has been transformed to an effective approach for LCA analysis and it is very user-friendly. The end-point damage rate estimation by this approach reveals the overall impacts of a system under three major indicators: human health, ecosystem, and resources.

Moreover, the CED method is used to measure fossil fuel-based energy consumption rates for all of the considered plants as this method has been widely utilized by various research groups to assess various types of fuel intake throughout the production process of a system [125]. It reflects the rate of power consumption from fossil fuels (oil, coal, and gas), biomass, renewables, and nuclear power. This approach is important to estimate the fossil fuel consumptions and substitute them with renewable sources. Fig. 5.6 highlights the life cycle impact assessment methods (Eco-indicator 99 and CED) used in this research.

Furthermore, the IPCC method is used to calculate the GHG emissions based on a 100-year time frame. This method not only considers emissions like carbon dioxide, methane, nitrogen oxide, and other GHGs, but also includes land transformation-based emission for impact analysis [152]. Finally, uncertainty analysis is carried out by the Eco-indicator 99 method to check the probability distribution of all the renewable power plants with respect to a number of single-score points. This method helps to examine the robustness of the measured impacts for this research.

5.3.4 Life cycle impact interpretation

The LCIs of the three renewable power plants incorporate crucial information regarding the distribution of the consumed energy and the material flow in building the plants, which are utilized to assess the environmental impacts systematically. The straightforward tendency appears that the main energy flow and impact occur during the fuel production stage and the raw material processing stage. By LCA we identify and compare the major factors that cause environmental hazards by solar-PV, pumped storage hydropower, and biomass power plants. A comprehensive LCA is carried out to compare which renewable power plant is a better choice in terms of the environment using the TRACI, Eco-indicator 99, CED, and IPCC methods. Furthermore, insightful recommendations are provided based on uncertainty analysis.

5.4. Results and interpretation

5.4.1 Environmental impact comparison

The overall inputs and outputs obtained by the RMF method are depicted in Table 5.4. The maximum input from nature is taken by the pumped storage hydropower plant, whereas the minimum input from nature is captured by the solar-PV plant. The outputs to air, water, and soil and solid wastes are highest for the pumped storage hydropower plant. On the other hand, the biomass plant has small outputs to water and air, which are lower than for the pumped storage hydropower plant. But the biomass power plant consumes the highest amount of inputs from nature. Overall, the solar-PV plant releases the lowest amounts of outputs to the air, water, and soil in comparison with pumped storage hydropower and biomass power plants.

The TRACI mid-point indicator-based method gives the comparative environmental impact assessment outcome for all three considered renewable power plants, using a cradle-to-gate analysis. After the systematic analysis, this method reveals environmental effects for 10 impact categories: eutrophication, acidification, smog, ecotoxicity, fossil fuel depletion, ozone depletion, global warming, respiratory effects, carcinogenics, and noncarcinogenics. Fig. 5.7 and Table 5.5 show the mid-point impact outcome

Table 5.4 LCA inputs and outputs of the considered plants using the RMF approach.

Impact category	Solar-PV	Pumped storage hydropower	Biomass	Unit
Inputs from nature	3.70E−06	2.66E−01	8.43E−01	kg
Outputs to air	6.09E−06	1.05	9.08E−01	kg
Outputs to water	1.17E−06	2.68E−03	3.14E−04	kg
Outputs to soil	6.07E−08	1.16E−04	3.94E−09	kg
Solid waste output	2.28E−08	1.63E−05	3.67E−08	kg

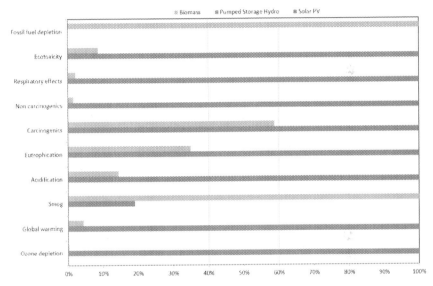

Figure 5.7 Normalized environmental impact outcomes, as determined using the TRACI mid-point approach.

for solar-PV, pumped storage hydropower, and biomass power plants per impact indicator by adding individual effects using the TRACI mid-point approach. The highest impact is set to 100%. The pumped storage hydropower plant exhibits maximum emission for all impact indicators except for fossil fuel depletion and smog, for which the biomass power plant exhibits the highest emission. The biomass power plant has little impact on ecotoxicity, respiratory effects, carcinogenics, noncarcinogenics, eutrophication, acidification, and global warming. The solar-PV power plant is the most environmentally friendly among the three plants according to the TRACI method, with the lowest impact on the environment.

Fig. 5.8 shows the LCA outcomes after weighting by the Eco-indicator 99 end-point approach. The results reveal that the solar-PV plant performs

Table 5.5 Mid-point environmental impacts of the considered plants as determined using the TRACI method.

Impact category	Solar-PV	Pumped storage hydropower	Biomass	Unit
Ozone depletion	6.81E−14	4.81E−08	8.28E−11	kg CFC-11-eq.
Global warming	4.35E−06	1.06	4.61E−02	kg CO_2-eq.
Smog	3.21E−07	4.62E−02	2.41E−01	kg O_3-eq.
Acidification	3.18E−08	5.78E−03	8.39E−04	kg SO_2-eq.
Eutrophication	1.08E−07	1.17E−04	4.11E−05	kg N-eq.
Carcinogenics	2.83E−13	2.52E−09	1.47E−09	CTUh
Noncarcinogenics	5.58E−12	2.218E−08	3.32E−10	CTUh
Respiratory effects	6.38E−10	3.45E−04	7.44E−06	kg PM2.5-eq.
Ecotoxicity	6.45E−06	7.45E−02	6.43E−03	CTUe
Fossil fuel depletion	6.73E−13	4.07E−07	4.81E−03	MJ surplus

in a much more environment-friendly manner than biomass and pumped storage hydropower plants. It is evident that the maximum effect is incurred by the biomass power plant for resources, while the minimum is obtained from the solar-PV plant for the three impact types. The pumped storage hydropower plant mainly impacts human health and ecosystem quality. The reason behind these impacts is the number of toxic materials present as constituents of the plant structure and its electricity production steps. Furthermore, the highest impact on ecosystems is made by the pumped storage hydropower plant (normalized value 7.74E−05), whereas the biomass power plant has the highest impact on resources. However, the solar-PV power plant impacts the ecosystem and resources 1.8E−03% and 1.34E−08%, respectively. Therefore, the Eco-indicator 99 end-point results reveal that the pumped storage hydropower plant is the most harmful and the solar-PV plant shows the lowest impact on our environment.

5.4.2 Fossil fuel-based energy consumption evaluation

The fossil fuel-based energy consumption rates at different stages of the solar-PV power plant, the pumped storage hydropower plant, and the biomass power plant are represented in Table 5.6. These values are obtained by the CED method, considering different types of fuel inputs like fossil fuels, renewables, nuclear power, and biomass energy throughout the

Environmental impact assessment of renewable power plants in the US

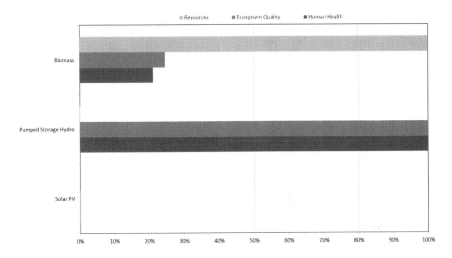

Figure 5.8 LCA outcome after weighting by the Eco-indicator 99 end-point approach.

Table 5.6 Fuel-based energy consumption rates of solar-PV, pumped storage hydropower, and biomass plants in the US.

Impact category	Solar-PV	Pumped storage hydropower	Biomass	Unit
Renewables	3.85	4.27E−01	4.20549E−05	MJ
Fossil fuels − oil	6.28E−07	8.35E−01	3.13E−02	MJ
Fossil fuels − gas	8.09E−06	2.73	1.78E−03	MJ
Fossil fuels − coal	3.72409E−05	7.61	1.04E−03	MJ
Biomass	1.34974E−06	1.50E−01	2.52929E−08	MJ
Nuclear	6.84384E−10	3.71	2.92E−04	MJ

plants' element manufacturing phases. The comparative evaluation of the total fuel input types in all of the plants' element production stages shows that the biomass power plant uses the lowest amount of fossil fuels compared to other plants and no renewables. Mostly oil is used in the biomass plant (3.13E−02 MJ). On the other hand, the solar-PV power plant uses most renewables (3.85 MJ). The pumped storage hydropower plant uses the highest amounts of fossil fuels and nuclear energy; mostly coal is consumed (7.61 MJ). Therefore, to reduce fossil fuel consumption, the solar-PV-based plant is the better choice of the three considered renewable power plants.

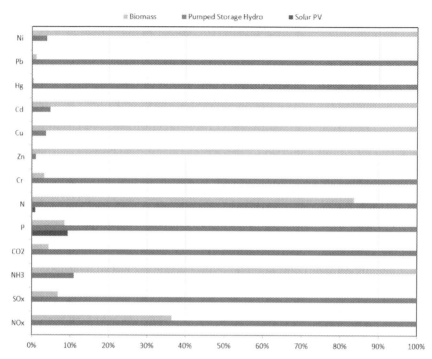

Figure 5.9 Metal- and gas-based emissions by renewable energy plants as determined using the Eco-points 97 method.

5.4.3 GHG emission factor estimation

Metal- and gas-based emission values by the renewable energy plants are depicted in Fig. 5.9 and Table 5.7. These values are obtained by the Eco-points 97 methodology. Maximum values are set to 100%. Among several categories, carbon dioxide (CO_2), nitrogen oxide (NO_x), sulfur oxide (SO_x), nitrogen (N), lead (Pb), mercury (Hg), chromium (Cr), and phosphorus (P) emission rates are highest in the pumped storage hydropower plant. The biomass-based power plant emits the highest amount of nickel (Ni), cadmium (Cd), copper (Cu), zinc (Zn), and ammonia (NH_3). The solar-PV plant emits minimal amounts of nitrogen (N) and phosphorus (P). Overall, the Eco-points 97 results show that the smallest amounts of metals and gases are emitted from the solar-PV power plant.

The GHG emissions by the renewable power plants are highlighted in Fig. 5.10 and Table 5.8. These values are obtained using the IPCC method. IPCC gives outcomes based on a 100-year time frame. The biomass-based power plant releases the lowest amounts of GHGs for most of the

Table 5.7 Plants' metal- and gas-based emissions as determined by the Eco-points 97 method.

Impact category	Solar-PV	Pumped storage hydropower	Biomass	Unit
NO_x	1.29E−05	1.86	6.79E−01	g
SO_x	4.45E−06	4.46	2.98E−01	g SO_2-eq.
NH_3	1.33E−06	3.83E−03	3.51E−02	g
CO_2	4.33E−03	1065.485271	46.55	g CO_2-eq.
P	3.28E−06	3.53E−05	2.99E−06	g
N	7.68E−05	8.43E−03	7.04E−03	g
Cr	2.38E−08	1.99E−04	6.48E−06	g
Zn	3.83E−11	5.21E−08	5.18E−06	g
Cu	9.88E−12	2.71E−08	7.38E−07	g
Cd	1.52E−12	5.06E−09	1.04E−07	g
Hg	2.38E−10	9.37E−07	3.06E−09	g
Pb	3.55E−09	1.34E−04	1.50E−06	g
Ni	1.51E−11	2.67E−08	7.01E−07	g

Table 5.8 GHG emissions as determined by the IPCC approach.

Impact category	Solar-PV	Pumped storage hydropower	Biomass	Unit
Carbon dioxide	4.03E−06	1.02	4.28E−02	kg CO_2
Methane	7.69E−09	1.24E−03	2.15E−05	kg CH_4
Nitrogen oxide	4.57E−10	3.16E−05	9.36E−06	kg N_2O
Land transformation	4.31E−14	6.07E−03	4.98E−07	kg CO_2-eq.
Other	1.54E−10	1.53E−04	1.12E−07	kg CO_2-eq.

categories, including carbon dioxide, methane, and land transformation, whereas the pumped storage hydropower plant releases the highest amounts of GHGs into the environment.

5.4.4 Comparison of impacts with other power plants

According to the mid-point and end-point indicator-based results analyzed by the TRACI and Eco-indicator 99 methods, respectively, electricity generated by a solar-PV plant shows greater environmental sustainability than electricity generated by other sources like lignite coal, natural gas, nuclear, biomass, and pumped storage hydropower. The LCA data for all of the plants located in the US are collected from the ecoinvent database. From the 10 major mid-point impact categories, electricity generated by coal has

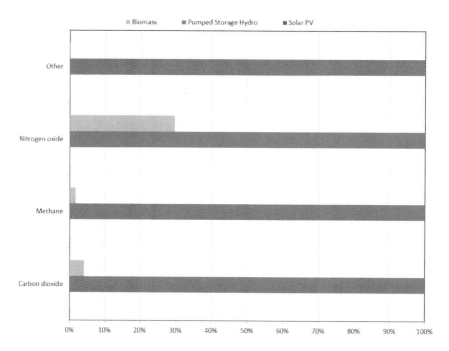

Figure 5.10 GHG emissions as determined using IPCC methodology.

the most detrimental effect on global warming, acidification, eutrophication, and respiratory effects (Table 5.9). However, for the carcinogenics, noncarcinogens, ecotoxicity, and fossil fuel depletion categories, electricity generated by natural gas has the most adverse effect. Energy produced by pumped storage hydropower plants affects ozone depletion more than other sources. Moreover, electricity generated from biomass has higher impacts on smog than other power generation sources. Lignite coal-based power plants are detrimental to human health and ecosystem quality (Table 5.10). Natural gas-based power plants are more harmful to resources than other power plants. Overall, solar-PV plants are best for human health, ecosystem quality, and resources, followed by nuclear power plants and biomass-based plants.

5.5. Uncertainty analysis

The results of the uncertainty analysis conducted for the three power plants are shown in Figs. 5.11 to 5.13. In the TRACI analysis outcome, the

Table 5.9 Mid-point impact comparison with other nonrenewable power plants.

Impact category	Solar-PV	Pumped storage hydropower	Biomass	Bituminous coal	Natural gas	Nuclear	Unit
Ozone depletion	6.81E−14	4.81E−08	8.28E−11	1.28E−09	6.38E−08	6.32E−08	kg CFC-11-eq.
Global warming	4.35E−06	1.06	4.61E−02	1.1	7.46E−01	1.03E−02	kg CO$_2$-eq.
Smog	3.21E−07	4.62E−02	2.41E−01	7.63E−02	1.68E−02	8.34E−04	kg O$_3$-eq.
Acidification	3.18E−08	5.78E−03	8.39E−4	9.35E−03	6.17E−03	7.32E−05	kg SO$_2$-eq.
Eutrophication	1.08E−07	1.17E−04	4.11E−05	1.99E−04	7.18E−05	6.01E−06	kg N-eq.
Carcinogenics	2.83E−13	2.52E−09	1.47E−09	8.77E−10	3.01E−09	2.18E−10	CTUh
Noncarcinogenics	5.58E−12	2.21E−08	3.32E−10	2.29E−08	3.91E−08	4.45E−10	CTUh
Respiratory effects	6.38E−10	3.45E−04	7.44E−06	4.54E−04	3.62E−04	6.08E−06	kg PM2.5-eq.
Ecotoxicity	6.45E−06	7.45E−02	6.43E−03	8.42E−02	9.60E−01	4.98E−03	CTUe
Fossil fuel depletion	6.73E−13	4.07E−07	4.81E−03	1.79E−01	1.87	1.04E−08	MJ surplus

Table 5.10 End-point impact comparison with other nonrenewable power plants.

Damage category	Solar-PV	Pumped storage hydropower	Biomass	Bituminous coal	Natural gas	Nuclear	Unit
Human health	6.58E−12	6.83E−07	1.45E−07	9.85E−07	5.58E−07	1.25E−08	DALY
Ecosystem quality	3.44E−07	1.91E−02	4.74E−03	2.61E−02	9.98E−03	7.47E−04	PDF★m^{2}yr
Resources	3.61E−13	2.18E−07	2.68E−03	9.81E−01	1.01	5.58E−09	MJ surplus

Environmental impact assessment of renewable power plants in the US

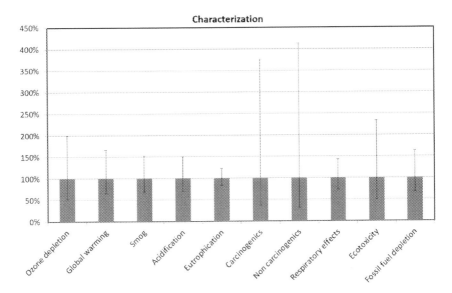

Figure 5.11 Probability distribution for the single-score impact category of the solar-PV power plant.

peak bars of the probability distributions for the impact categories reveal the greatest possibilities of uncertainty. The probability distribution of the solar-PV power plant (Fig. 5.11) shows more uncertainty in the carcinogenics, noncarcinogenics, and ecotoxicity categories. The lowest amount of uncertainty for the solar-PV plant is shown in eutrophication and acidification.

The uncertainty analysis results of the pumped storage hydropower plant are shown in Fig. 5.12. The highest level of uncertainty is shown in the carcinogenics, ecotoxicity, and eutrophication categories. The lowest amount of uncertainty is shown in the global warming, acidification, and respiratory effects categories. The uncertainty analysis results of the biomass power plant are shown in Fig. 5.13. The highest level of uncertainty is shown in the noncarcinogenics, ecotoxicity, eutrophication, and ozone depletion categories. Other categories show minimal amounts of uncertainty. Overall, the high numbers of small bars for all three cases reveal that the environmental performance evaluation by the TRACI method for all indicators is robust regarding uncertainties.

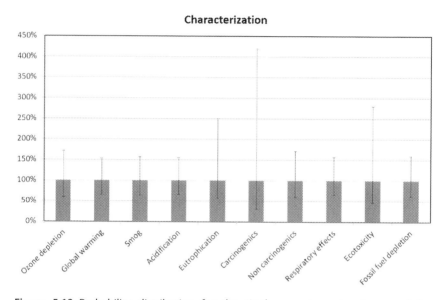

Figure 5.12 Probability distribution for the single-score impact category of the pumped storage hydropower plant.

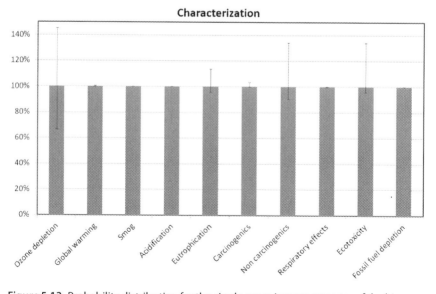

Figure 5.13 Probability distribution for the single-score impact category of the biomass power plant.

5.6. Sensitivity analysis

Sensitivity analysis using the Impact 2002+ approach has only been conducted for the pumped storage hydropower plant as it shows higher impacts than other renewable plants in the US. Energy consumption from the high-voltage electricity mix is found to be responsible for the highest impact for the pumped storage hydropower plants. Therefore, seven high-voltage electricity mix scenarios from various countries are considered for the sensitivity analysis. The cases are as follows:
- Case 1 – high-voltage electricity mix of the US;
- Case 2 – high-voltage electricity mix of France;
- Case 3 – high-voltage electricity mix of Switzerland;
- Case 4 – high-voltage electricity mix of Belgium;
- Case 5 – high-voltage electricity mix of Denmark;
- Case 6 – high-voltage electricity mix of the Czech Republic;
- Case 7 – high-voltage electricity mix of Brazil.

The case studies are compared based on the variation in high-voltage electricity mix of these countries used in the pumped storage hydropower plants. According to the analysis results presented in Table 5.11, the electricity mix of Brazil has the highest impact. The electricity mix of France has the lowest impact for the categories of carcinogens (2.70E−10 DALY), respiratory organics (8.65E−12 DALY), respiratory inorganics (2.10E−08 DALY), climate change (7.64E−09 DALY), ozone layer (7.27E−13 DALY), and land use (2.63E−04 PDF★m^2yr) than other countries. Moreover, the electricity mix of Brazil has the lowest impact for the categories of radiation (9.11E−11 DALY), ecotoxicity (1.42E−03 PDF★m^2yr), acidification (3.55E−04 PDF★m^2yr), and fossil fuels (5.94E−09 MJ surplus) of all mixes considered. However, the US high-voltage electricity mix has the lowest impact for radiation (5.71E−11 DALY) and a medium impact for other categories in pumped storage hydropower production. The reason behind these diverse results is evident after analyzing their inventory datasets. The inventory datasets show that the high-voltage electricity mix of Brazil emits the largest amounts of waste heat. This is the main reason for the high environmental impacts associated with high-voltage electricity mix. Similarly, for the US waste heat emissions from high-voltage electricity production are higher than for France, Belgium, and Denmark due to the increased waste heat emission.

Table 5.11 Sensitivity analysis of seven pumped storage hydropower plants in various countries.

Impact category	Unit	United States	France	Switzerland	Belgium	Denmark	Czech Republic	Brazil
Carcinogens	DALY	1.38E−09	2.70E−10	3.21E−10	5.64E−10	5.83E−10	8.59E−10	1.06E−08
Respiratory organics	DALY	4.51E−11	8.65E−12	9.71E−12	3.62E−11	3.14E−11	1.81E−11	2.43E−11
Respiratory inorganics	DALY	1.25E−07	2.10E−08	2.38E−08	4.53E−08	4.29E−08	1.04E−07	1.05E−08
Climate change	DALY	6.21E−08	7.64E−09	1.03E−08	2.80E−08	5.15E−08	6.51E−08	1.71E−08
Radiation	DALY	5.71E−11	3.51E−09	2.08E−09	1.88E−09	5.47E−10	9.64E−10	9.11E−11
Ozone layer	DALY	8.06E−12	7.27E−13	4.98E−12	4.85E−12	6.56E−12	2.62E−12	1.57E−12
Ecotoxicity	PDF* m²yr	7.12E−03	2.93E−03	2.69E−03	3.24E−03	2.67E−03	3.59E−03	1.42E−03
Acidification	PDF* m²yr	4.20E−03	6.16E−04	6.12E−04	1.39E−03	1.37E−03	3.46E−03	3.55E−04
Land use	PDF* m²yr	4.06E−04	2.63E−04	2.94E−04	4.47E−04	5.24E−04	4.82E−04	2.19E−03
Fossil fuels	MJ surplus	1.02E−07	6.89E−09	1.55E−08	1.13E−08	1.68E−07	2.68E−09	5.94E−09

5.7. Discussion

The selection of a power plant in a country is largely dependent on the availability of resources and reliability, as these change depending on the geographic position. This chapter compares solar-PV, pumped storage hydropower, and biomass plants in the US. A previous research conducted by Hou et al. [107] showed that a solar-PV power plant in China emits 60.1 g CO_2-eq./kWh, whereas Varun et al. [58] revealed that a similar kind of plant located in Japan emits about 53.4 g CO_2-eq./kWh. This study shows that a solar-PV power plant in the US emits approximately 4.03 g CO_2-eq./kWh. The availability of resources and small transportation distance make solar-PV plants more environment-friendly. Research by the Chinese Academy of Engineering revealed that a run-of-river hydropower plant in China emits 5.5 g CO_2-eq./kWh [168]. A US-based pumped storage hydropower plant emits much more into the environment (1020 g CO_2), because it consumes fossil fuel to pump water to the storage. However, a World Nuclear Association report of 2015 showed that a run-of-river hydropower plant emits about 5.5 g CO_2-eq./kWh [171], while Varun et al. reported a rate of 237 g CO_2-eq./kWh. They also reported that a China-based biomass production plant emitted about 35 g CO_2-eq./kWh into the environment. But this study shows that a biomass plant located in the US emits approximately 42.8 g CO_2-eq./kWh. Therefore, the location of a plant is an essential factor in the overall sustainable production of energy, as raw material extraction, transportation, and construction are crucial parts in impact assessment and changes based on plant position. The outcome of this study was compared with existing studies highlighted in Fig. 5.14.

In comparison to conventional fossil fuel-based power generation units and nuclear plants, RET-based plants could have a much reduced environmental impact for all of the assessment indicators. But biomass plants have a high impact in the smog category, higher than that of nuclear plants. Researchers need to identify ways of abating this impact of biomass plants to make energy production cleaner in the future.

For estimation of the GHG emission of the considered plants, the IPCC method has been used. However, there might be a small deviation in outcome using different methods. The time period for impact estimation by the IPCC approach is taken as 100 years. Besides, the outcome may vary for different time frames.

TRACI has been used for mid-point impact assessment due to its acceptability for US-based sustainability evaluation. But this method does not

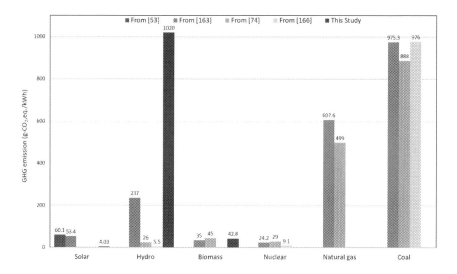

Figure 5.14 Comparison of the findings with existing studies.

provide the impact incurred by radiation, whereas a few global methods of LCA, like Eco-indicator 99, ILCD, etc., highlight the associated effects of radiation for any system. Therefore, the main constraints of this research are summed up as follows:

- LCA of other renewable energy resources which are used in the US is not considered here due to a lack of suitable datasets.
- The substitution of harmful materials and machines in the context of the environment without compromising electricity outcomes and efficiencies of the plants has not been investigated.
- Means to abate the use of fossil fuels by the plants have not been studied.
- Cost–benefit analysis for the considered plants has not been conducted.

The further extension of this research should be an overcoming of the specified constraints.

5.8. Conclusion

In conclusion, this study highlights the environmental impacts of three different renewable power plants, namely solar-PV, biomass, and pumped storage hydropower plants, through a systematic LCA analysis by the TRACI, Eco-indicator 99, CED, Eco-points 97, RMF, and IPCC methods, using SimaPro software. The results reveal that the solar-PV plant is the most environmentally sustainable as it emits less carbon dioxide

(4.03E−06 kg CO_2-eq./kWh), methane (1.62E−07 kg CO_2-eq./kWh), and nitrogen oxide (1.42E−07 kg CO_2-eq./kWh) than others. Moreover, the total environmental impacts for all the end-point indicators, like human health, ecosystem quality, and resources, are lower for solar-PV plants than for others. The pumped storage hydropower plant is a threat to the environment based on its high environmental impact as it consumes fossil fuel or nuclear energy to pump water to the storage, whereas biomass power plants cause medium damage to the environment except for smog and ozone layer depletion. However, comparison of the impacts of solar-PV, biomass, pumped storage hydropower, nuclear, natural gas, and bituminous coal plants reveals that bituminous coal has the strongest effect on global warming (1.1 kg CO_2-eq./kWh) and eutrophication (1.99E−04 kg N-eq./kWh), natural gas has the strongest effect on ozone depletion (6.38E−08 kg CFC-11-eq./kWh), carcinogenics (3.01E−09 CTUh/kWh), ecotoxicity (9.60E−01 CTUe/kWh), and fossil fuel depletion (9.60E−01 MJ surplus/kWh), and biomass has the highest effect on smog (2.41E−01 kg O_3-eq./kWh). However, solar-PV plants exhibit the smallest effects for all impact indicators. The outcomes from this research will guide governments and investors in taking appropriate decisions by installing the best renewable power plant considering environmental aspects and available resources in the US. Overall, future research should be concentrated on investigating the effects and avoiding the environmental hazards of installed power plants.

CHAPTER SIX

Comparative environmental impact assessment of solar-PV, wind, biomass, and hydropower plants

Electricity generation technologies harness existing renewable or nonrenewable energy sources to produce electricity. Nonrenewable electricity generation systems use resources like coal or diesel, whereas renewable systems use wind energy, solar energy, municipal waste incineration, biomass, or hydropower. The choice of a power plant in different nations depends on the availability, abundance, and reliability of required resources. It is evident that renewable energy generation resources provide a more sustainable solution than fossil fuels. However, as complete systems, renewable energy generation systems also have impacts on humankind and ecosystems. The main objective of this chapter is to addresses the environmental effects caused by different types of renewable energy generation systems through process-based life cycle impact analysis. A comparative study of different renewable energy generation technologies is carried out, including wind, photovoltaic (PV), biomass, and hydropower. These resources vary among nations based on their geographic position, latitude–longitude, and resources; this chapter thus also compares electricity generation systems in Europe, North America, and Oceania, considering the fact that Europe has enormous PV and hydropower installations, while the US has notable biomass plants and Oceania has wind power plants. Life cycle impact analysis has been carried out by the Institute of Environmental Sciences (CML), Eco-indicator 99, cumulative energy demand (CED), and Eco-points 97 approaches. The superiority of this work lies in the creation of a comprehensive life cycle inventory (LCI) using a reliable global database, assessment of the impacts for about 10 mid-point impact categories and 3 end-point indicators, and inclusion of the fossil fuel-based energy consumption rate of each plant. The key findings reveal that PV power plants have the highest environmental impacts in the categories of ozone layer depletion, fresh water aquatic ecotoxicity, and marine aquatic ecotoxicity, while biomass plants impact the categories of abiotic depletion, global warming, photochemical oxidation, acidification, and eutrophication. Moreover, wind power plants have the strongest environmental effect on human toxicity and terrestrial ecotoxicity. Overall, hydropower plants are found to be much more environment-friendly than other renewable electricity generation systems.

6.1. Introduction

With the growth of the world population and the economy, the demand for energy has increased enormously. To meet this rising demand, fossil fuels are burned at power plants, which produce hazardous chemicals and pollute our environment [100]. Globally, about 2 million pounds of CO_2 are released into the air every second by fossil fuel-based power plants [30]. People are becoming more conscious of the adverse ecological impact of conventional power generation units. To reduce this pollution, in the last few decades renewable energy technologies (RETs) have become popular [28,29,97,104,156]. Moreover, RETs produce electricity with enhanced cost-effectiveness, increased efficiency, and superior environmental profiles [93,95,96,157]. Most studies on RETs have focused on productivity enhancement and cost–benefit analyses [33,38]. Very little research has been carried out to quantify the environmental effect of RETs, as it is considered small [24,109,148]. However, renewable power plants are associated with emissions [15,53,94,107,159], mostly during the extraction of raw materials, production of elements from the extracted raw material, transportation of the materials to the plant, and the end-of-life waste management period [48,160]. These emissions from RETs are less than those from conventional power generating units, but not negligible [58,163].

Now the question arises how beneficial the shift from conventional to renewable-based energy generation would be. Energy optimization-based models are not able to answer this question. In a life cycle assessment (LCA)-based study carried out by Siddiqui et al., the environmental impacts of nuclear, wind, and hydropower plants in Ontario were compared, raising problems like variations in LCIs and modeling approaches and therefore a lack of a complete depiction of the total effects [56]. Furthermore, they have not considered the solar and biomass power plants and the research is only applicable to Ontario, as global data have not been used. Another comparative LCA-based environmental effect assessment of solar-PV and wind energy systems was done by Nugent et al., which is not very methodologically rigorous [53]. They only measured and compared

GHG emissions; other impacts were not considered. Some previous studies only investigated the effects of one element of the overall plant like polymer solar cells [74] or PV panels [25] and focused on reducing the amount of hazardous raw material used to minimize the impact. A comprehensive LCI was not included in either of the abovementioned LCA analyses, so they lack LCA modeling accuracy and reliability. Suh et al. used hybrid approaches to select the system boundaries of LCIs [177]. Pomponi et al. [178], Wiedmann et al. [179], and Lenzen et al. [180,181] showed that hybrid LCA is likely to yield more accurate findings than process-based LCA. Also, they showed that even simple process systems could provide greater dominant eigenvalues based on real data. LCA and sustainability analysis aid decision making [182,183]. The uncertainties in LCA should be abated by compliance to a standardized approach utilizing input–output-based hybrid LCA methods [184,185]. The analytical LCA approach reveals a good estimation of the output uncertainty [186]. The innovative approach of using national average input–output data assists in reducing gaps in traditional LCA inventories, such as hybrid analysis [187]. It provides an understanding of the net energy savings that are feasible with a well-made energy system. Country-wise LCA of different renewable production technologies was carried out by various researchers who have not considered global databases like ecoinvent in life cycle input–output measurement [15,27,32,41,48,57,160,166,167,188]. A separate LCA research was conducted on PV generation systems [102,103], wind turbines [165], hydropower plants [134], and biomass power plants [51] by different research groups, but none quantified the amount of fossil fuel-based energy consumption by the considered plants during construction, usage, and end-of-life management. Therefore, more research is needed in this field to find the exact emission rate from each production step of the plant elements and to discover ways to reduce those emissions. To the authors' knowledge, until now no research has been conducted to assess and compare the environmental impacts of four RETs, namely solar power plants, wind power plants, hydropower plants, and biomass power plants. For that reason, this research aims to elucidate the effects of these plants on our environment by a dynamic LCA approach. LCA is a useful tool to quantify the environmental impact of any product, unit, or system utilizing different methods considering the material flows, input resources, and output emissions [59,120,170].

The ecoinvent database is utilized to collect the material flows and gather all the life cycle input and output data for creating a new LCI and system boundary for all of the considered plants. LCA is carried out using SimaPro software by the CML, Eco-indicator 99, CED, and Eco-points 97 methods. In brief, the key contributions of this research work are as follows:

- The mid-point (cradle-to-gate) and end-point (cradle-to-grave) environmental impacts of four different RET-based power plants, including solar power plants, wind power plants, hydropower plants, and biomass power plants, are assessed and compared.
- Hazardous metal and greenhouse gas (GHG) emissions from the considered plants over their lifetime are evaluated.
- The fossil fuel-based energy consumption during the manufacturing of plant elements, the installation and operation of the plants, and waste management is estimated.
- Uncertainty analysis is conducted for all considered cases.

This research is superior to previous LCA-based environmental effect assessments of renewable energy systems in several ways. For example, a comprehensive LCI is created following a reliable global database, the impacts for about 11 effect indicators are assessed, including GHG emissions, and the fossil fuel-based energy consumption rate of each plant is taken into account. Considering all of these aspects, this will be the first research that highlights and compares the total environmental impacts of solar, wind, biomass, and hydropower plants.

In light of the above, the rest of the chapter is organized as follows. Section 6.2 highlights the materials and methods to carry out a systematic LCA for assessing environmental profiles, calculating GHG emission factors, and conducting uncertainty analysis. The results and discussion are presented in Section 6.3, in four parts – comparison of plants' ecological effects, evaluation of their fossil fuel-based energy consumption, estimation of their GHG release, and uncertainty analysis. Finally, concluding comments on the research outcome are given in Section 6.4.

6.2. Materials and methods

The main focus of this work is to assess the atmospheric threats of four different renewable power plants, i.e., solar, wind, biomass, and hydropower plants, and compare their ecological consequences based on 11 impact indicators. To fulfill this goal a systematic LCA approach is utilized. LCA is considered to be the most powerful tool for assessing the environmental effects of any product, unit, or system [111,113]. The application of LCA includes, but is not limited to, impact assessment, uncertainty analysis, and sustainability investigation [114–117]. This LCA approach is usually accomplished by following standards 14040:2006 and 14044:2006 of the International Standardization Organization (ISO) [118,119]. The LCA software SimaPro is utilized to assess the environmental effects. LCA is performed by collecting the inputs and outputs at each manufacturing step of the element, and ecological impacts are calculated utilizing four fundamental steps:

1. goal and scope definition for assessment, where the aim is defined and the LCA boundaries are set (ISO 14040);
2. LCI analysis, where inputs and outputs at every step of the manufacturing processes are assembled (ISO 14041);
3. life cycle environmental impact evaluation following the data compiled at the previous stage, where yield emissions and input assets are divided into their specific effect indicators and changed into the same unit for comparative evaluation (ISO 14042);
4. impact outcome interpretation and suggestions to actualize the goals of this study (ISO 14043).

In the following subsections, the above LCA steps are described briefly to depict the LCA methodology which is followed in this research.

6.2.1 Goal and scope definition

The primary goal of this LCA is to calculate the environmental threats of four different renewable energy generation systems, namely a solar power plant, a wind power plant, a biomass power plant, and a hydropower plant. This LCA also compares their impact outcomes to identify the best renewable electricity generation plant. The LCA is conducted from both cradle-to-gate (from raw material extraction to the factory gate) and cradle-to-grave (from resource extraction to disposal) perspectives for all considered systems [117,120]. Therefore, the overall LCA considers the complete life cycle of the plants, including raw material extraction and transportation to the production plant, plant element manufacturing and transportation to the plant, plant installation and operation, and end-of-life recycling and disposal of elements. Figs. 6.1 to 6.4 illustrate the absolute and relative amounts of energy and material flow (in MJ) in different stages of the solar power plant, the wind power plant, the hydropower plant, and the biomass power plant to produce unit power. The functional unit is 1 kWh, which represents the amount of energy produced by the overall system for which

Table 6.1 Data sources for the considered renewable power plants.

Plant	Data source
Solar power plant	Electricity, production mix photovoltaic, at plant/AT U/AusSD U
Wind power plant	Electricity, at wind power plant 2 MW, offshore/OCE U/AusSD U
Hydro power plant	Electricity, hydropower, at power plant/BE U/AusSD U
Biomass power plant	Electricity, biomass, at power plant/US U

the impacts are assessed [111,113]. The arrows indicate the material flow directions. Different colors highlight specific types of material flows and the percentages demonstrate the weighted amounts of respective materials.

6.2.2 Life cycle inventory

This second step of LCA concerns the input raw materials, input energies, input resources, output emissions, end-of-life recycling quantities, and disposal amounts per unit process considered in the LCA system boundary. For getting all these rates, it is required to collect the LCI data from any recognized database, industry, or literature review. The inventories for various renewable energy plants change among countries based on the location and available resources. Europe has gigantic PV and hydropower assets, whereas the US has more biomass availability and Oceania has many windy locations. Therefore, when collecting data from the renowned ecoinvent database [121,122] we considered solar and hydropower plants in Europe, a wind power plant in Australia, and a biomass power plant in America. Table 6.1 shows the data source for each of the considered plants in this research. Basically all renewable power generation units follow almost the same steps throughout their lifetime, including extraction of raw materials from mines, transportation of the extracted raw materials to the industry, manufacturing of plant elements from the raw materials, transportation of raw materials to the plant area, installation of the plant, production of electricity, and end-of-life waste management. For that reason, we consider a common system boundary for all four plants in our LCA analysis. Fig. 6.5 highlights the system boundary considered in this impact assessment method. The overall inputs of a plant are raw materials, water, nonrenewable energy, organic and inorganic chemicals, transportation, and other available resources at the specific plant location. On the other hand, the outputs are 1 kWh of electricity and waste emissions.

Comparative environmental impact assessment of renewable power plants

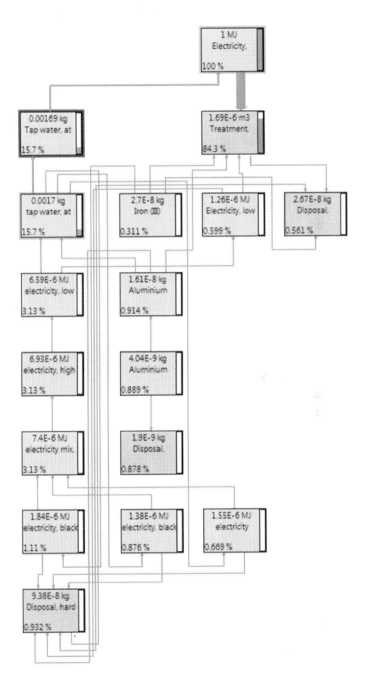

Figure 6.1 Material flow sheet for 1 MJ of solar energy generation.

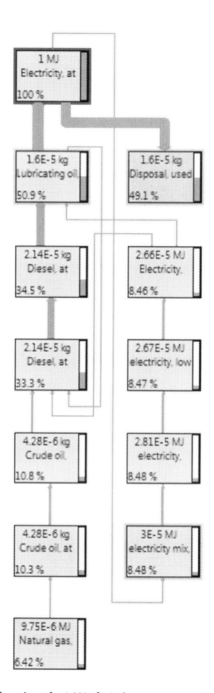

Figure 6.2 Material flow sheet for 1 MJ of wind energy generation.

Comparative environmental impact assessment of renewable power plants 143

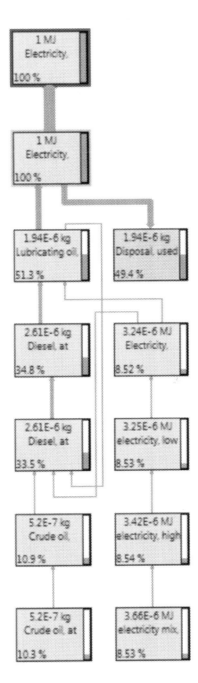

Figure 6.3 Material flow sheet for 1 MJ of hydro energy generation.

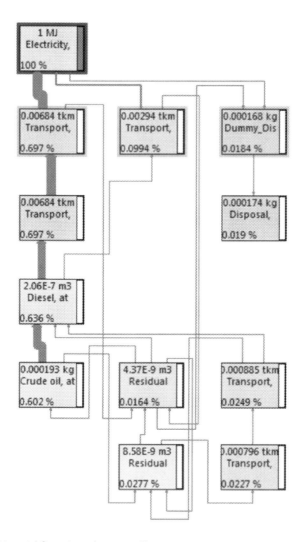

Figure 6.4 Material flow sheet for 1 MJ of biomass energy generation.

6.2.3 Life cycle impact assessment

LCA was carried out by the systematic steps of creating a system boundary for solar, wind, hydro, and biomass power plants considering LCIs, assessing impacts by different approaches, and comparing the outcomes among the four considered plants. To form the LCI, the ecoinvent database was used to gather the data for the plants [101,121,122]. The ecoinvent database contains international industrial LCI data for raw material extrac-

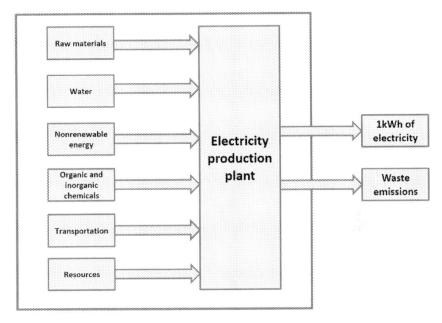

Figure 6.5 Common system boundary for all power generation processes used in this LCA analysis.

tion, transportation, energy usage, etc. All data for the considered plants were collected from the ecoinvent database as it is an internationally well-known and reliable data source. We created the LCI for the same amount of reference flow and nonfunctional units (1 kWh). SimaPro software was used in assessing the impacts after creating the LCI and system boundary [138]. The CML method is utilized to determine the cradle-to-gate environmental impacts for 10 impact indicators, as it is based on the problem-oriented (mid-point) approach. It provides potential ecological effects under obligatory and additional impact categories such as eutrophication, acidification, toxicity, and global warming [189]. The CED method is used to measure fossil fuel-based energy consumption rates for all considered plants as this method has been widely utilized by various research groups to assess various types of fuel intakes throughout the production process of a system [125]. This method considers all types of energy usage such as nuclear, biomass, renewable, and fossil fuel (oil, coal, and gas) during the overall life cycle of the plants and provides a breakdown of energy consumption. It is pivotal to get information on the fossil fuel consumption by the plant to replace it by renewable energy and improve the environmental profile. The method

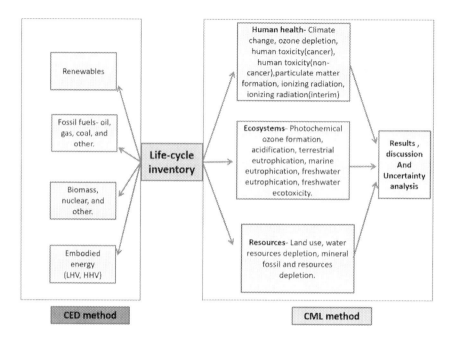

Figure 6.6 Life cycle impact assessment methods used in this research.

also provides a total sum of energy based on both low and high heating values. Fig. 6.6 highlights the life cycle impact assessment methods (CML and CED) used in this research. Furthermore, the Intergovernmental Panel on Climate Change (IPCC) method is used to estimate the GHG emissions based on a 100-year time frame. This method not only considers emissions like carbon dioxide, methane, nitrous oxide, and other GHGs, but also includes land transformation-based emission for impact analysis [152]. This approach offers three benefits in assessing GHG emissions: (i) it ensures optimal use of the data source in a comprehensive manner, (ii) it establishes transparency in the assessment, and (iii) it provides insights for policymakers into climate solutions. Finally, uncertainty analysis is carried out by the Eco-indicator 99 method to check the probability distribution of all the renewable power plants with respect to a number of single-score points. This method is advantageous as it includes a weighting approach in LCA. After weighting, it enables a single score to be assigned for each product or process, calculated based on the relative environmental impact. This method helps to examine the robustness of the measured impacts for this research.

6.2.4 Life cycle impact interpretation

The LCIs of the four renewable power plants incorporate crucial information regarding the distribution of the consumed energy and the material flow in building the plants, which are utilized to assess the environmental impacts systematically. The straightforward tendency appears that the main energy flow and impact occur during the raw material processing to create the basic parts of the plants such as PV panels, wind and hydro turbines, etc. By LCA we identify and compare the major factors that account for environmental hazards by solar, wind, hydro, and biomass power plants. A comprehensive LCA is carried out to determine which renewable power plant is a better choice in terms of the environment using the CML, Eco-indicator 99, CED, and IPCC methods. Furthermore, insightful recommendations are provided based on sensitivity analysis.

6.3. Results and discussion

6.3.1 Comparison of environmental impacts

The overall input and output scenarios are assessed and compared among all the considered renewable power plants by the Raw Material Flows (RMF) approach. This method tracks the total mass flow of raw material and emission by summing all elementary flows contained in the ecoinvent database. In theory, if all inventories are mass-balanced, the sum of the inflows should equal the total amount of outflow, but this is rarely the case as combustion incorporating oxygen input is often not included and moisture balances are also not always tracked. Fig. 6.7 depicts the outcome obtained by the RMF method. The highest input from nature is taken by the biomass power plant, whereas the lowest input is captured by the hydropower plant. The outputs to water and soil and solid waste are highest for the solar power plant. However, the wind power plant provides about half the outputs to water and land compared to the solar power plant. Overall, the hydropower plant releases the lowest amounts of output into the air, water, and soil.

The CML mid-point method gives the environmental impact outcome for all four considered renewable power plants based on a cradle-to-gate analysis. After the systematic assessment, this method reveals the environmental effects in 10 impact categories, including eutrophication, acidification, photochemical oxidation, terrestrial ecotoxicity, marine aquatic ecotoxicity, human toxicity, ozone layer depletion, global warming, and abiotic depletion. Fig. 6.8 shows the similar mid-point impact outcomes

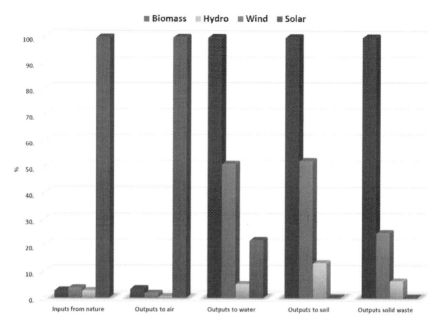

Figure 6.7 Comparative LCA inputs and outputs of the considered renewable energy plants using the RMF method.

for solar, wind, hydro, and biomass power plants per impact indicator by adding individual effects using the CML mid-point approach. The highest impact is set to 100%. The biomass power plant has the highest impact for 5 of 10 impact indicators (abiotic depletion, global warming, photochemical oxidation, acidification, and eutrophication). Furthermore, the wind power plant has the highest impact for two impact indicators (human toxicity and terrestrial ecotoxicity). However, impacts in categories like ozone layer depletion, freshwater ecotoxicity, and marine aquatic ecotoxicity are highest for the solar power plant. Overall, the hydropower plant releases the lowest amounts of hazardous materials in all 10 impact categories.

Fig. 6.9 shows the LCA outcomes after weighing by the Eco-indicator 99 end-point approach. The results reveal that the hydropower plant performs better environmentally than the other considered plants. It is evident that for all three end-point impact categories (human health, ecosystem quality, and resources), the highest impact is observed for the biomass power plant and the lowest impact is observed for the hydropower plant. The reasons behind these impacts lie in the toxic materials used as constituents of the plant structure and its electricity production steps. Furthermore, the

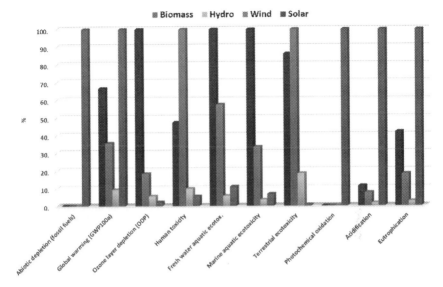

Figure 6.8 Comparison per impact indicator using CML mid-point methodology by adding individual effects. The highest impact is set to 100%.

impact of the hydropower plant on ecosystems is highest (100%), whereas the impacts of the wind power plant, solar power plant, and hydropower plant are 75%, 62%, and 40%, respectively. The biomass power plant has the highest impact (100%) on human health; the impacts of the hydro, wind, and solar power plants are only 5%, 15%, and 35%, respectively. Therefore, the Eco-indicator 99 end-point results reveal that the biomass power plant is the most harmful and the hydropower plant is the least hazardous to our environment.

6.3.2 Fossil fuel-based energy consumption evaluation

The fossil fuel-based energy consumption rates at different stages of the solar power plant, the wind power plant, the hydropower plant, and the biomass power plant are represented in Fig. 6.10, as determined by the CED method, considering different types of fuel inputs like fossil fuels, renewables, nuclear energy, biomass, and embodied energy throughout the plant elements' manufacture phases [125]. The comparative evaluation of the total fuel input types in the plant elements' production shows that the wind power plant uses most embodied energy in case of both higher heating values (HHVs) and lower heating values (LHVs) as it consumes a considerable amount of embodied energy during wind turbine manufacturing

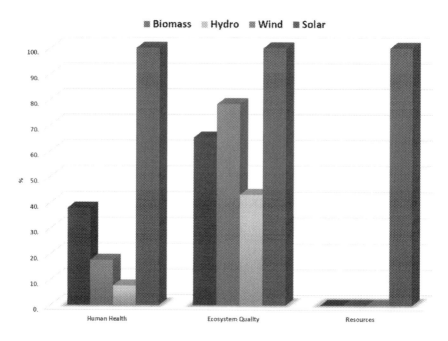

Figure 6.9 LCA outcomes after weighting by the Eco-indicator 99 end-point approach.

[165]. The biomass power plant uses the lowest amounts of fossil fuels and renewables. On the other hand, the solar power plant uses most fossil fuels as it consumes a considerable amount of carbon-based energy during PV cell production [24]. The hydropower plant uses higher amounts of renewable and embodied energies; however, it uses less fossil fuel and nuclear energy. Therefore, due to the low fossil fuel usage, the hydropower plant is the better choice of the four considered renewable power plants.

6.3.3 GHG emission factor estimation

The metal- and gas-based emissions by the renewable energy plants are compared in Table 6.2, which is obtained by the Eco-points 97 methodology, a systematic approach for assessing metal and hazardous gas emissions by any system [115]. The maximum amount is set to 100%. Carbon dioxide (CO_2), nitrous oxide (NO_x), sulfur oxide (SO_x), ammonia (NH_3), nitrogen (N), copper (Cu), cadmium (Cd), and Nickel (Ni) emission rates are highest in the biomass power plant. The solar power plant emits most nitrate, pesticide, waste, phosphorus (P), and dissolved organic carbon (COD) into

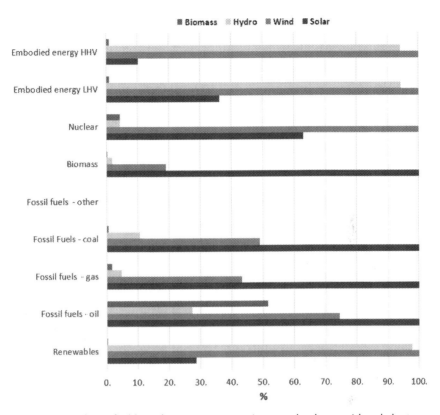

Figure 6.10 Relative fuel-based energy consumption rates by the considered plants, as determined using the CED method.

the environment. The wind power plant emits most zinc (Zn), mercury (Hg), and chromium (Cr). Overall, the Eco-points 97 results show that the smallest amounts of metals and gases are emitted from the hydropower plant.

The GHG emissions by the chosen renewable power plants, as determined using the IPCC method, are compared in Fig. 6.11. The IPCC method is a logical approach for assessing climate change and economic aspects in countries of the United Nations [152]. It gives outcomes based on a 100-year time frame. The solar power plant releases most GHGs for most of the categories like carbon dioxide, methane, and land transformation, whereas the hydropower plant releases the least GHGs into the environment.

Table 6.2 Metal- and gas-based emissions by renewable energy plants, as determined using the Eco-points 97 method.

Label	Solar	Wind	Hydro	Biomass
NO_x	10.9193	5.8056	2.2319	100
SO_x	14.2976	10.5791	1.3819	100
NMVOC	0.1287	0.0379	0.0115	100
NH_3	3.4999	1.5367	0.305	100
Dust PM10	61.9885	100	73.8542	1.844
CO_2	65.5707	35.4815	9.3409	100
Ozone layer	100	69.9877	1.6657	0.3384
Pb (air)	100	43.0902	3.1933	0.0123
Cd (air)	100	41.3672	1.286	0.0383
Zn (air)	74.2694	100	12.2064	0.0148
Hg (air)	28.203	100	19.8198	0.1098
COD	100	37.1593	12.5098	4.5535
P	100	7.445	0.7095	1.0776
N	62.1776	38.7468	0.5074	100
Cr (water)	41.2868	100	16.6704	1.4833
Zn (water)	3.259	0.841	0.2166	100
Cu (water)	5.9322	1.653	0.4033	100
Cd (water)	5.9287	1.8405	0.4436	100
Hg (water)	100	91.9914	10.3636	75.3376
Pb (water)	100	31.9876	5.3378	67.6722
Ni (water)	9.7473	2.531	0.64	100
AOX (water)	100	0.912	0.0457	0.0004
Nitrate (soil)	100	16.0244	0.7336	0.0072
Metals (soil)	100	27.7947	7.6282	0.0544
Pesticide soil	100	6.5552	1.1609	0.0092
Energy	0.9247	0.2094	100	0.9057

6.3.4 Uncertainty analysis

The uncertainty analysis outcomes of the four considered renewable energy-based power plants are shown in Figs. 6.12 to 6.15. This analysis is carried out following the Eco-indicator 99 approach. The peak bars of the probability distributions for the single-score impact category of the plants reveal the greatest possibility of uncertainty. The probability distribution of the PV power plant (Fig. 6.12) shows more uncertainty in environmental impact calculations based on the higher number of peak bars in the prob-

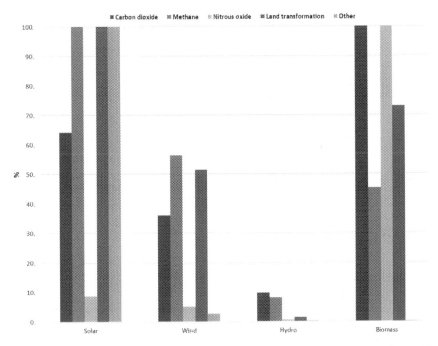

Figure 6.11 Relative GHG emissions by the plants, as determined using the IPCC methodology.

ability distribution to the single-score points of the *x*-axis. On the other hand, the smaller number of peak bars in the probability distribution to the single-score points (Pt of the *x*-axis) obtained for the biomass power plant (Fig. 6.15) demonstrates the smaller uncertainty in environmental impact calculations. The wind and hydropower plant probability distributions depicted in Figs. 6.13 and 6.14, respectively, demonstrate medium uncertainties in the environmental impact calculations for single-score points. The high number of small bars for all four cases reveals that environmental performance evaluation by the Eco-indicator 99 method for all indicators is robust regarding uncertainties.

6.3.5 Comparison with other studies

The environmental impacts of solar-PV, wind, hydro, and biomass plants for different mid-point indicators are collected from prior studies [2,10, 172–174,190]. A summary of the results is depicted in Table 6.3. It is evident that the maximum impact of the wind power plant is observed for the ionizing radiation category and the highest effect of the biomass power

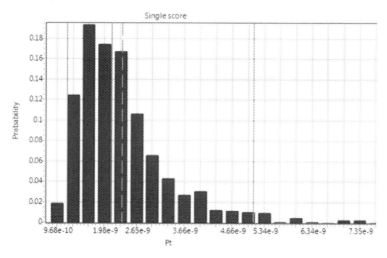

Figure 6.12 Probability distribution for the single-score impact category of the PV power plant.

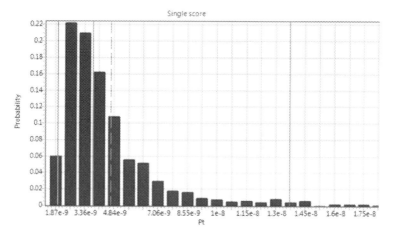

Figure 6.13 Probability distribution for the single-score impact category of the wind power plant.

plant is observed for the global warming category. Solar plants are mostly responsible for terrestrial ecotoxicity and hydropower plants have lower impacts in all mid-point impact categories. The outcomes from this research will guide the industries and communities in taking appropriate decisions by installing the best renewable power plant considering the environmental aspects and available resources.

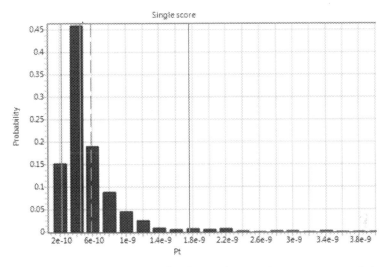

Figure 6.14 Probability distribution for the single-score impact category of the hydropower plant.

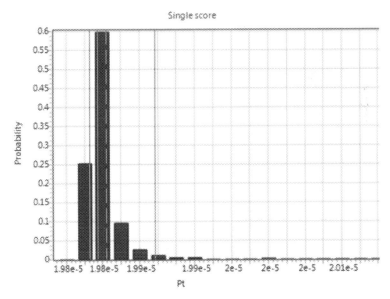

Figure 6.15 Probability distribution for the single-score impact category of the biomass power plant.

The relative GHG emissions of various power plants obtained from previous studies [2,10,173,190] are summarized and depicted in Fig. 6.16. GHG emission by coal power plants is higher than that by renewable

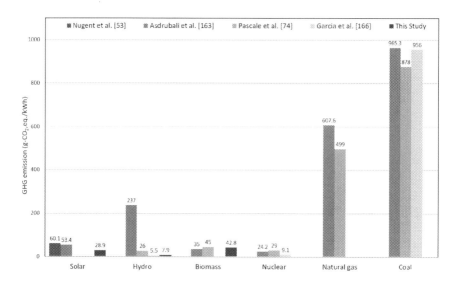

Figure 6.16 Outcome comparisons with prior studies.

plants with a high rate of 97%. However, GHG emission of hydropower plants is lowest according to most of the prior literature. The impact of biomass plants on global warming is highest among the considered renewables [2,10]. It is necessary to abate these damages to the environment, and future research should be directed toward identifying the highest impacting materials in the renewable plant's life cycle. Based on the findings of this study, it is evident that the following recommendations should be considered to achieve greener, cleaner, and more sustainable electricity generation:

- It is important to replace fossil fuel-based electricity consumption by renewables. The appropriate selection of renewable resources depends on the geographic location of the plant.
- It is required to develop a global LCA approach for each renewable power plant, which would be valid irrespective of the geographic location.
- It is necessary to develop a universal database focusing on RET, which should consider data for all types of renewable energy systems all over the world.
- It is recommended to identify the most hazardous elements in plants and to replace them with environmentally superior alternatives.
- It is essential to recycle the waste to optimize improvement.

Table 6.3 Important findings and comparison with other studies.

Ref.	Topic	Considered plants	Major impact categories from the power plants	Important findings
[179]	Application of hybrid life cycle approaches to emerging energy technologies – the case of wind power in the UK	Wind	GHG emissions	Compared to integrated hybrid LCA, the IO-based hybrid LCA approach is easier to implement. It requires less effort to compile the model framework because only input–output matrices are required, no process and upstream information.
[38]	LCA of electricity generation technologies: overview, comparability, and limitations	Hard coal, lignite, natural gas, oil, nuclear, biomass, hydro-electric, solar-PV, and wind	Emissions of GHG, NO_x, and SO_2	The cycling emissions are responsible for about 7% of CO_2, NO_x, and SO_2 emissions in the considered plant.
[185]	Hybrid LCA of algal biofuel production	Biomass	Carbon dioxide emissions	Improvements in the algae to improve the biocrude conversion process would greatly reduce the energy costs, and thus increase the net carbon savings.

continued on next page

Table 6.3 (continued)

Ref.	Topic	Considered plants	Major impact categories from the power plants	Important findings
[51]	LCA applied to electricity generation from renewable biomass	Biomass	Acidification, eutrophication, photo-oxidant formation	The impacts of 2-MW and 10-MW biomass plants are compared through an LCA analysis.
[53]	Assessing the life cycle GHG emissions from solar-PV and wind energy: a critical metasurvey	Solar-PV and wind	Climate change, respiratory effects	The physical features of the solar systems are mostly responsible for harmful emissions.
[54]	Life cycle sustainability assessment of key electricity generation systems in Portugal	Solar-PV, wind, hydro, coal, and natural gas	Global warming potential, acidification potential, primary energy demand	Hydropower is the best option from the environmental and socioeconomic perspectives.
[55]	Integrated life cycle sustainability assessment of the Greek interconnected electricity system	Solar-PV, wind, hydro, biogas, and lignite	Photo-chemical oxidation potential, ozone depletion potential	The solar-PV plant is the most sustainable form of electricity from a social perspective. The wind power plant is the most sustainable form of electricity from environmental and economic perspectives.

continued on next page

Table 6.3 (continued)

Ref.	Topic	Considered plants	Major impact categories from the power plants	Important findings
[56]	Comparative assessment of the environmental impacts of nuclear, wind, and hydroelectric power plants in Ontario: an LCA	Nuclear, wind, and hydro	Acidification potential, global warming potential	Among nuclear, hydro, and wind-based power plants, wind plants are mostly responsible for environmental hazards in the categories of acidification, eutrophication, and human toxicity potential.
[10]	LCA of the environmental impacts of solar-PV and solar-thermal systems	Solar-PV and solar-thermal	Human toxicity, energy demand	The environmental impact of a solar-PV plant is higher than that of a solar-thermal system of the same capacity.
–	This study	Solar, wind, hydro, and biomass	Global warming, ozone layer depletion, terrestrial ecotoxicity, eutrophication	PV power plants have the highest environmental impacts in the categories of ozone layer depletion, fresh water aquatic ecotoxicity, and marine aquatic ecotoxicity, while biomass plants impact abiotic depletion, global warming, photochemical oxidation, acidification, and eutrophication.

6.4. Conclusion

In conclusion, we compared the environmental impacts of four different renewable power plants, namely solar, wind, biomass, and hydropower plants, through a systematic LCA analysis by the CML, Eco-indicator 99, Eco-points 97, RMF, and IPCC methods, using the ecoinvent database and SimaPro software. The results reveal that the hydropower plant is the least damaging to the environment as it emits less CO_2, NO_x, and SO_x than others. Moreover, the total environmental impacts for all of the end-point indicators, like human health, ecosystem quality, and resources, are lower for hydropower plants than for others. The biomass power plant has the highest impact on the environment in several impact assessment categories, whereas solar and wind power plants cause medium damage to the environment. The outcomes from this research will guide the government and investors in taking appropriate decisions by installing the best renewable power plant considering environmental aspects and available resources. The replacement of environmentally hazardous elements and/or instruments without reducing electric output performance of the power plants has not been examined. Overall, to reduce the environmental hazards of an installed power plant, future research should be concentrated on investigating the effects and reducing those impacts.

CHAPTER SEVEN

Advanced energy-sharing framework for robust control and optimal economic operation of an islanded microgrid system

Energy sharing through a microgrid (MG) is essential for islanded communities to maximize the use of distributed energy resources and battery energy storage systems (BESSs). Proper energy management and control strategies of such MGs can offer revenue to prosumers (active consumers with distributed energy resources) by routing excess energy to their neighbors and maintaining grid constraints at the same time. This chapter proposes an advanced power-routing framework for a solar-photovoltaic (PV)-based islanded MG with a central storage system (CSS). After optimization of the economic operation of the MG, the power routing and energy sharing in the MG are determined in the day-ahead stage. A modified droop controller-based real-time control strategy has been established that maintains the voltage constraints of the MG. The proposed power-routing framework is verified via a case study for a typical islanded MG. The outcome of the optimal economic operation and a controller verification of the proposed framework are presented to demonstrate the effectiveness of the proposed power-routing framework. Results reveal that the proposed framework performs a stable control operation and provides a profit of 57 AU$/day at optimal conditions.

7.1. Introduction

Renewable energy technologies (RETs) are becoming progressively more popular for both off-grid and interconnected-grid utilization. Of all the RETs, solar-PV is the favored and most promising one due to its efficiency, availability, low cost, low maintenance requirements, reliability, robustness, and clean production [10,11,100,198]. The inconsistency of energy production using PV panels highlights the need for integration of a BESS in the local grid or MG to achieve hybrid operation [191–195]. The combination of PV panels, energy storage, and local loads forms a conventional MG system [196]. These MGs play a crucial role in satisfying the load demand of isolated localities, such as rural villages, islands, enclaves, and exclaves, by smart energy sharing with cost-efficient operation [95,112,197], smooth control, and optimized management [145,198,199]. In addition, the surplus power from the MGs can be routed to a neighbor's home when the load demand of the PV owners is fulfilled and the central storage reaches its charge limit. Thus, the energy can be utilized by a simple method of local power routing, which offers both revenue for the owners and much-needed power for the neighbors [100,200,201]. Consequently, from both a technical and an operational point of view, the MG, due to its small size, short transmission distance, easy maintenance, robust control, and economic operation, can be a good option for energy sharing.

In islanded operation of an MG framework, the main aim is to reduce the generation, operation, and maintenance costs and system losses during sharing to optimize the prosumers' revenue. This can be achieved by a systematic priority formation through various optimization techniques. Nonlinear programming (NLP) is a favored approach to maximize the profit while satisfying the operational constraints of the storage system. It has been utilized in some recent studies to achieve required objectives such as load-sharing optimization [202], best-case scenario development for energy generation scheduling [48], optimal load scheduling [203], and off-grid power management hierarchical structure formation [204]. None of the previous optimization-based studies focused on maximization of the profits of energy trading from a stakeholder's viewpoint of an MG framework through maximum power routing to consumers while fulfilling the constraints of the CSS.

On the other hand, an incompetent control mechanism of voltage regulation during energy sharing is the foremost hindrance to the advancement of MG frameworks [11,205]. The limitations associated with renewable resources, like low inertia and storage requirements, make the control of the proposed islanded MG more challenging [206]. Additional complexity arises when load sharing happens between PV owners and powerless neighbors in the MG framework to ensure overall power balancing [144,207], state of charge (SoC) regulation [208], and voltage balancing [209]. Researchers have proposed several decentralized control schemes for the smooth operation of islanded MGs with distributed generation and battery energy storage [93,210,211]. A frequency-based power management system is highlighted in [212], but it is not applicable for frameworks with isolated BESSs. Another similar method proposed by Dan et al. in [213] is also not compatible with systems with an isolated BESS, while in [214] the technique is updated by applying the droop control approach for a particular solar/BESS-based hybrid structure. Moreover, regulating the voltage and frequency level of an MG bus is a way of power stabilization during energy routing [104,215]. Power balancing in an MG framework can also be achieved by utilizing the previously proposed charging and discharging algorithm [216] and using the PI controller of previous work [217] to regulate the deviation between the DC bus and the central storage currents. However, in [207] a unique architecture of power management is proposed which is applicable for regulating power flow in an MG. The major drawback of this approach is that it is not possible to apply in current frameworks, as it requires real-time data to adjust the power routing unit.

Recently, Fernandez et al. developed a game-theoretic energy-sharing model to provide cost savings for a smart neighborhood in Sydney [218]. Their Nash game theory-based model helps to reduce the energy cost by using central storage during the peak time instead of consuming energy from the main grid; about 9.17% and 9.68% more cost saving was achieved in summer and winter, respectively, compared with the noncooperative method. However, in this research all revenue is considered as cost savings of prosumers, as they sell excess energy to consumers. Moreover, no prosumers are competing among each other at NLP-based profit optimization, presented in this chapter. Another recent work by Nizami et al. has considered revenue sharing among all consumers [219], whereas this research has taken into account the profit maximization from a prosumer's perspective. Moreover, this research provides stable control operation of the proposed framework. The contribution by Akter et al. provided an energy-sharing model for a residential MG considering hierarchical energy management for three different cases following the participation of PV units and BESSs [220]. Their work dealt with a rule-based mechanism for intelligent energy management. However, this work offers a state-of-the-art approach through NLP-based optimal economic operation of an MG framework for the maximum profit of prosumers and robust control operation of the proposed system.

The main focus of this work is to ensure maximum utilization of the available solar power and battery storage resources in a community. This chapter proposes a novel energy-sharing framework for a remote locality where an MG is the only means of meeting the prosumers' load demand and routing excess energy to their neighbors to fulfill their energy demands. In this chapter, an NLP-based optimization model is presented for the day-ahead scheduling of maximum available power routing after fulfilling prosumers' and consumers' load demands, as well as the constraints of CSSs for an off-grid remote MG framework. In the NLP formation, the optimization focuses on boosting the PV owners' revenue by maximum power routing to neighboring consumers after meeting their own load demands. Optimal economic operation of the proposed framework has been achieved and applied to improve the MG's effectiveness. Furthermore, the power-routing framework is verified to be a stable control operation strategy with proper voltage regulation utilizing a droop controller in the power control loop of the inverter. Therefore, the key contributions of this chapter can be summarized as follows:

1. A smart load-sharing framework for an off-grid remote community is developed. This framework provides cost-effective performance for smart houses as it routes surplus energy to nearby traditional houses without wasting it.
2. An NLP-based optimization model is developed and applied for the day-ahead scheduling of excess power routing for an optimal profit to stakeholders using the proposed MG framework.
3. A novel off-grid power-routing management strategy is proposed that measures the instant load demands and PV generation and routes the desired power to specific loads through priority formation utilizing local and central controllers, while the prosumers are provided the highest priority.
4. The proposed MG framework is verified by a droop controller through a real-time voltage control method, which provides stable DC and AC bus voltages and ensures better performance of the grid.

In the light of the above, the remainder of the chapter is organized as follows. The power-routing framework is introduced in **Section 7.2**. The optimization-based energy-sharing model is discussed in **Section 7.3**. **Section 7.4** depicts the power-routing control strategy of the proposed framework. The results to validate the profit optimization and controller performance and a discussion are presented in **Section 7.5**. Concluding remarks on the research outcomes are presented in **Section 7.6**.

7.2. Power-routing framework

The power-routing framework follows the concept of a solar home system, where small-scale commercial loads (such as a poultry farm) are interconnected with the own consumptive load [206]. Thus, the energy remaining after meeting the household load demand is utilized through an autonomous off-grid arrangement. In an MG framework, a number of low-capacity generators and BESSs are usually interconnected to a supply chain that provides electricity to several clients, where energy sharing takes place among suppliers and consumers. A benefit of this situation is that homes in an island or rural area where no central grid is available can share energy using the MG framework. Moreover, a group of 10–20 houses in an urban zone can interconnect and share extra renewable energy among themselves by this kind of system [221], in which the MGs may be connected to or disconnected from the central grid for a time based on demand. This can reduce the pressure on the central grid. Furthermore, it is economical as the shared energy rate is cheaper than grid power [39]. However, the MG power quality is required to be maintained properly, following its distinctive features, control strategies, and power management system. In the proposed system, a small cluster of houses within a short radius are interconnected to share electricity through the islanded off-grid arrangement. A few of them have a PV power generation facility. The surplus power in the MG is routed to the neighboring powerless houses after meeting the demands of prosumers by this framework.

Fig. 7.1 represents the conceptual architecture of the power-routing framework in a remote off-grid locality. The houses with PV systems are considered as prosumers and represented by smart houses, whereas a traditional house represents a consumer without any PV facility. The PV generation units of smart houses are connected to a DC bus via DC/DC boost converters and the power generated from the PV systems is stored in a CSS. The outputs of the CSSs are connected to the MG controller (MGC), which controls the MG inverter. The MGC and LC controls are based on hierarchical control strategies for effective power routing and stable operation. The main task of the MGC is to gather the power requirements measured by the local measuring units and route the desired power to specific loads. Generator and traditional houses are connected to the AC bus. The generator is utilized to supply power in an emergency situation when PV generation and the CSS capacity are not enough to fulfill the energy demands of prosumers and consumers. The overall power-routing process is represented in Fig. 7.2. Prosumers are prioritized during power dispatch. After fulfilling their total demand the remaining power is routed to neighboring traditional houses of the cluster according to the energy-sharing optimization model. For a square deal, a uniform distribution is used among the traditional houses. All of them follow a prepaid payment method based

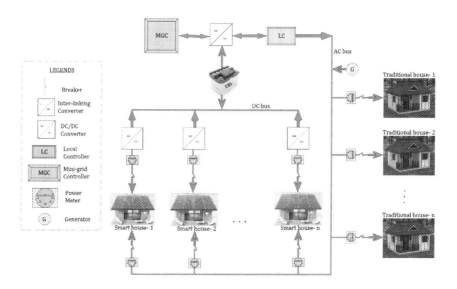

Figure 7.1 The conceptual architecture for the power-routing framework.

on a unit energy price in buying electricity from the grid, and the total revenue is distributed among the eight prosumers by curtailing their consumption cost. The power meter plays a vital role in power flow calculations, payment information sharing, and management. The optimization model for the energy sharing is described in Section 7.3. A modified control strategy is used with droop control that satisfies the voltage constraints for the energy sharing. The control strategy is discussed in Section 7.4.

7.3. Optimization-based energy-sharing model

The main idea is to develop a power-routing framework that allows prosumers to utilize their excess PV generation by trading it with traditional consumers. Therefore, it will only sell energy to a consumer if there is any excess power in the MG. However, in an emergency when there is not enough power from both the PV panels and the CSS, the prosumers and consumers will draw energy from the generator. As the objective is to develop optimal power routing from a prosumers' perspective, only the profits of prosumers are considered. Generator activity is neglected as it does not have any role in profit maximization from PV generation. However, this case study is aimed at showing that the PV generation and CSS size are more than enough to fulfill the prosumers' demand and route the excess

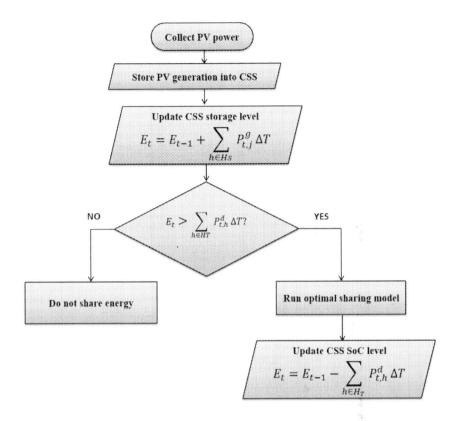

Figure 7.2 Power routing management strategy.

power. To ensure efficacy of the proposed model, it checks how much excess energy would be available to route for making optimal profit for prosumers after fulfilling their power demand and the battery-related constraints. The NLP optimization considers day-ahead forecasts of demand and PV generation to determine the power trading.

To fulfill the main objective of the energy-sharing model, maximizing the profits of energy trading while satisfying the operational constraints of the CSS, the sharing model is formulated as a nonlinear optimization problem:

$$\max_{P^s_{t,h}, \lambda_t} \sum_{t \in T, h \in H} P^s_{t,h} \Delta T \lambda_t \tag{7.1}$$

subject to

$$P^s_{t,h} = P^d_{t,h} \quad \forall h \in H_S, \tag{7.2}$$

$$0 \leq P^s_{t,h} \leq P^d_{t,h} \quad \forall h \in H_T, \tag{7.3}$$

$$\sum_{h \in H_T} P^s_{t,h} \Delta T \leq SoC_t E_{cap} + \sum_{j \in G} P^g_{t,j} \Delta T \quad \forall t, \tag{7.4}$$

$$SoC_{min} \leq SoC_t \leq SoC_{max}, \tag{7.5}$$

$$\lambda_{min} \leq \lambda_t \leq \lambda_{max}, \tag{7.6}$$

$$SoC_t = SoC_{t-1} + \frac{\sum_{j \in G} P^g_{t,j} \eta_{ch} - \frac{\sum_{h \in H} P^s_{t,h}}{\eta_{dis}}}{E_{cap}}. \tag{7.7}$$

Here, the objective function of (7.1) maximizes the profits from energy trading. The optimization variables $P^s_{t,h}$ represent the power supplied to house $h \in H$ during period $t \in T$ and λ_t indicates the per-unit cost for traded energy during $t \in T$. The duration of the time period is represented by ΔT. The load demand of houses $h \in H$ during period $t \in T$ is represented by $P^d_{t,h}$ and Eqs. (7.2)–(7.3) indicate that the energy demand of smart houses $h \in H_s$ has the maximum priority in the energy-sharing model, whereas energy is traded with traditional houses $h \in H_T$ only after supplying the smart house demand. As the objective is to develop optimal power routing from a prosumer's perspective, only the profits for prosumers are considered.

The energy balance constraints in (7.4) and (7.5) ensure that energy is traded with traditional houses $h \in H_T$ only when there is excess energy stored in the CSS or if there is excess PV generation. Here, $P^g_{t,j}$ represents the power generation of the PV system $j \in G$ during $t \in T$. SoC_t indicates the SoC of the CSS during $t \in T$ and E_{cap} is the maximum storage capacity of the CSS. Eqs. (7.5) and (7.6) indicate the upper and lower bounds for SoC_t and λ_t, respectively. The linear dynamics of the CSS SoC is indicated in (7.7). Here, η_{ch} and η_{dis} indicate the charging and discharging efficiencies of the CSS.

7.4. Power-routing control strategy

The MG power-routing strategy follows an ordered control operation of the DC-to-DC converter and DC-to-AC inverter control to obtain a stable performance throughout the period. The overall control method for effective energy sharing is described below.

7.4.1 DC-to-DC converter control

Here, the power supplied to each house ($P_{t,j}^s$) is determined from the optimization model discussed in **Section 7.3**. Depending on the shared information, a control signal is sent either to discharge the CSS to supply a house load or to shed some or all of the house load. A PI controller is utilized following previous work in [217] to find the difference between the DC bus and CSS currents. Overall, the active power of the DC bus is regulated using the designed DC-to-DC converter controller.

7.4.2 DC-to-AC inverter control

This converter control maintains the output voltage of the DC bus at the CSS. The $P - V_{DC}$ droop-based control method is utilized to balance the active and reactive powers of the system. The total amount of PV generation and the total load supplied by the CSS at any instant are calculated as follows:

$$P_{total,t}^g = \sum_{j \in G} P_{t,j}^g, \tag{7.8}$$

$$P_{total,t}^s = \sum_{i \in H_T} P_{t,j}^s + \sum_{i \in H_S} P_{t,j}^s. \tag{7.9}$$

Multiple houses share their power to achieve the MG operation. As a result, to ensure robust power sharing among multiple houses, with stable voltage and frequency regulation in the MG, a droop control method is adopted in the power control loop of the inverter. The conventional P/f and Q/V droop control schemes are adopted to achieve the control objective. The droop control schemes can be expressed as follows [222]:

$$\begin{aligned} \omega &= \omega^* - D_P(P - P^*), \\ V &= V^* - D_Q(Q - Q^*), \end{aligned} \tag{7.10}$$

where $\omega = 2\pi f$, where f is the frequency of the system, V^*, ω^*, P^*, and Q^* are reference values for voltage, angular velocity, active power, and reactive power, respectively, and D_P and D_Q are the droop coefficients. The inverter controller structure is illustrated in Fig. 7.3. The AC and DC buses shown in Fig. 7.4 are interfaced through four-quadrant inverters and LCL filters. The filtered current (i_o^{abc}), the voltage (v_o^{abc}), and the output current (i_l^{abc}) in an ABC reference frame are fed into an ABC-DQO converter, which results in their corresponding DQO components. The synchronization is

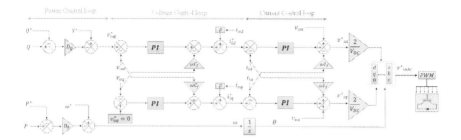

Figure 7.3 Inverter control structure.

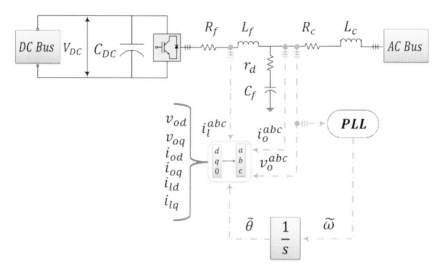

Figure 7.4 Inverter circuit diagram with LCL filter.

carried out utilizing a phase-locked loop, which generates the necessary synchronizing angle for the ABC-DQO converter. Overall, the inverter controller includes a power control loop, a voltage control loop, and a current control loop. The control parameters are given in Table 7.1. The power control loop is designed with a conventional droop control scheme that maps the output power from each house with the droop coefficients and with the optimization-based energy-sharing unit. A decoupled control with a feed-forward technique is adopted for the voltage and current control loops. The main objective of the voltage control loop is to generate reference values for i_d and i_q, which are i_d^* and i_q^*, respectively. The current control loop generates the necessary reference signal for the associated pulse-width modulator which controls the inverter. The detailed config-

Table 7.1 Control system parameters.

Parameter	Value
Nominal AC bus RMS voltage	240 V
Nominal DC bus voltage	700 V
Damping resistor, r_d	2×10^{-3} Ω
L_f and C_f	250×10^{-6} H, 0.0169 F
ω^*	314.16 rad/s
Power controller	
D_P and D_Q	3.14×10^{-5} rad/s/W
	1.36×10^{-4} V/VAR
Voltage controller	
k_p and k_i	7 and 800
Current controller	
k_p and k_i	0.3 and 20
DC/DC converter	
k_p and k_i	0.2 and 1.1

uration of the inverter controller along with its small-signal model can be found in [223] and [224].

7.5. Simulation and results

7.5.1 Simulation setup

The proposed method has been run through a simulation case study considering an MG framework in a remote islanded area of Bangladesh (Kutubdia) having an area of about 1000 m² where no electricity facility exists. A total of 20 houses are considered, out of which 8 are considered to be smart houses with PV generation units. The overall test model has a CSS with a maximum charging capacity of 90%, eight DC/DC converters having an output of 700 V DC, and a DC/AC converter (240 V). Fig. 7.5 reveals the average daily aggregated load of the prosumers and consumers used for the case study. Figs. 7.6 and 7.7 show the individual load demands of all 8 prosumers and 12 consumers, respectively. The electricity demands of both prosumers and consumers are maximum at night time and minimum at noon. The peak individual and aggregated loads of prosumers are about 6 kW and 20 kW, respectively, whereas the maximum individual and aggregated load demands of consumers are considered as approximately 6.5 kW and 24 kW, respectively. The aggregated load demand of consumers

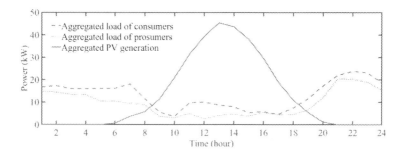

Figure 7.5 Aggregated PV generation and load profile of prosumers and consumers.

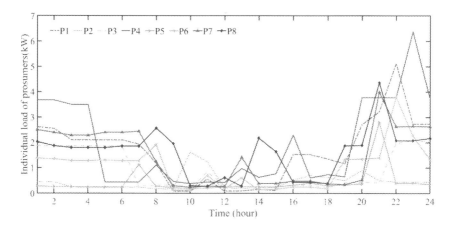

Figure 7.6 Individual load profiles of prosumers.

is higher than for prosumers as the number of consumers is 1.5 times that of prosumers in the proposed MG framework.

The solar irradiation profile of the location is depicted in Fig. 7.8. At midday, the maximum irradiation value of about 0.8 kW/m^2 is achieved. The irradiation value is zero until 5 a.m., then increases gradually until midday, and falls steadily after that time. At night no solar production is possible due to zero irradiation, although the maximum loads appear at night time for both prosumers and consumers. Therefore, the considered CSS with a capacity of 180 kWh has to play a vital role in energy storage and supply during night time, prioritizing the prosumers' load demands.

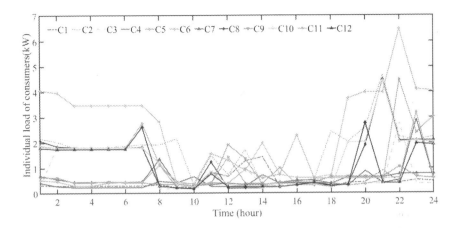

Figure 7.7 Individual load profiles of consumers.

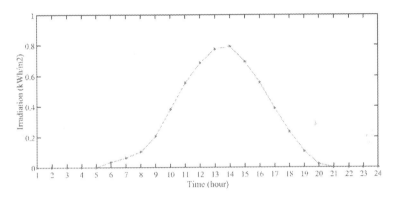

Figure 7.8 Solar irradiation profile at the MG location.

7.5.2 Results of the energy-sharing model

Upon running the simulation, the nonlinear problem (7.1)–(7.7) of **Section 7.3** has been optimized. The objective function is optimized with respect to the considered constraints using MATLAB®. Figs. 7.9 and 7.10 show the power demand and supply from the CSS for all the smart houses and traditional houses of the community, which are obtained from the simulation operation of the energy-sharing model. It is evident that all of the prosumers' hourly load demands are fulfilled. But all the hourly loads of the consumers could not be supplied by the CSS of the MG framework due to the unavailability of energy in the CSS, especially in the late night and early morning hours. At these times, when PV generation and CSS storage

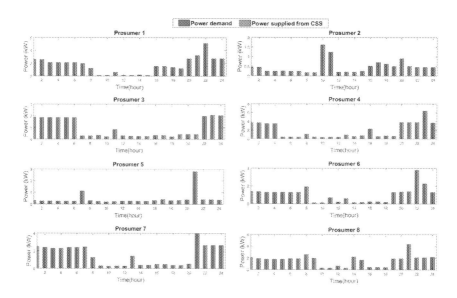

Figure 7.9 Hourly prosumers' demand and supply status.

are not enough to route power to the consumers, the demand would be supplied by the generator. As the objective is to develop optimal power routing from a prosumer's perspective, only the profits of prosumers are considered. Generator activity is neglected as it does not have any role in profit maximization by the available routed power, the main objective of this study. However, this case study ensures that on a typical day there is excess energy and prosumers make some profit by trading it. The SoC profile of the CSS is highlighted in Fig. 7.11. The minimum and maximum charge limits of the CSS are maintained as 10% and 90% over the whole day. At the beginning of the day, the charge level was about 53%, which reduced gradually until 10 a.m. to supply load demands with zero PV power generation in the morning. Because of high PV penetration at 10 a.m. to 7 p.m. the storage increases rapidly to 90%, and after that it gradually falls down to 33% at midnight as the CSS supplies power to the consumers' and prosumers' houses.

The hourly profit of the MG framework due to excess energy routing to the consumers is depicted in Fig. 7.12. The profit evaluations reveal that the greatest revenue rate occurred around midday, as the CSS got the highest storage at that time due to peak PV generation. The profit is zero for the first 9 hours of the day, as the CSS is unable to supply power due its small storage and zero PV production. At night time, a moderate profit

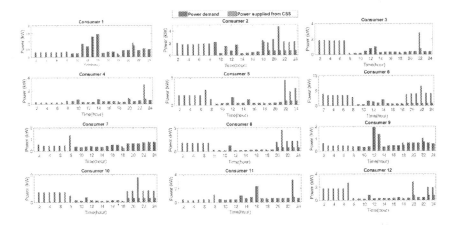

Figure 7.10 Hourly consumers' demand and supply status.

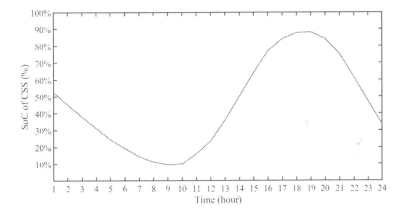

Figure 7.11 %SoC of the CSS.

with an average hourly rate of 3 AU$ is achieved by stakeholders through this proposed NLP-based optimal power-sharing framework. The average rate of trade is considered as 0.5231 AU$. The amount of total daily excess energy for routing is found to be approximately 109.57 kWh, which would have been wasted if not shared. Therefore, the total revenue of a typical day is about 57.32 AU$ (20,635.2 AU$/yr) with the proposed framework.

7.5.3 Controller operation outcomes

The proposed method is evaluated in simulation-based case studies. Real-life irradiation values are used for the PV generation. The output power

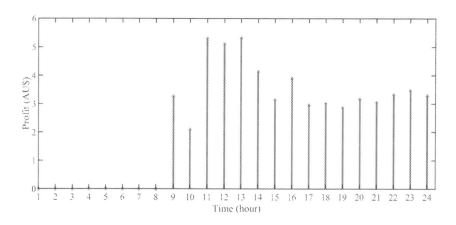

Figure 7.12 Hourly profit from the MG framework.

generated from the 8 PV panels of the prosumers is utilized to meet their load demand and then any excess energy is routed to 12 neighboring loads. The robustness of the designed controller is tested on an hourly basis assuming that the inverter provides similar behavior at every hour. During the moment of energy dispatch the inverter should provide stable operation, which is checked by the controller. The AC and DC loads of each house are connected to individual AC and DC buses of the MG. As a result, to ensure stable operation of the MG during various power dispatch schedules the control operation is carried out, and thus both the AC and DC bus voltages are regulated. The bus voltage profiles of the proposed method of controller operation for a selected time period are depicted in Fig. 7.13. It is observed that both the AC and DC bus voltages were quite stable throughout the period except for initial fluctuations. The community storage plays a vital role in regulating both bus voltages. In particular, during this control operation, the community storage acts as a slack bus that can support the DC bus voltage through the DC/DC converter. The AC bus voltage is regulated using four-quadrant operation of the interconnecting inverter. As both AC and DC loads are supplied accordingly by the PV and the community storage, both bus voltages are well regulated.

The acceptable frequency operating range for the continuous uninterrupted MG operation is about 50 Hz. It is shown in Fig. 7.14 that the frequency at the AC side of the MG is regulated within the operating range for diverse optimal power scheduling due to the robust nature of the droop-based inverter controller.

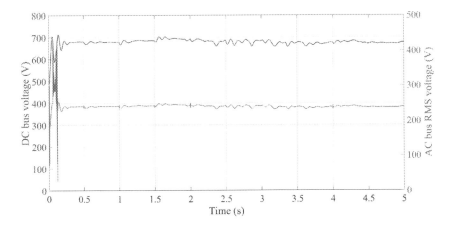

Figure 7.13 AC and DC bus voltages.

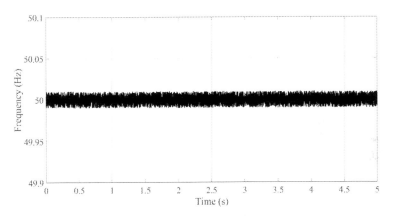

Figure 7.14 AC bus frequency.

7.6. Conclusion

In this chapter, a new power-sharing approach by a PV/CSS hybrid source-based remote area MG framework is proposed. This method performs all the necessary tasks such as load sharing among 8 PV owner units and 12 neighbor units, battery charging and discharging, and PV power curtailment with proper communication among the units. To ensure the efficacy of the MG system, we (i) developed a novel energy-sharing framework, (ii) optimized this based on profit maximization for cost-efficient operation, and (iii) verified that a typical droop controller-based control method gave stable functioning. This research work is novel in developing an NLP-based optimization model for day-ahead scheduling of available power routing using the proposed smart power-routing framework, which routes any excess energy to neighbor loads and thereby utilizes energy that would otherwise have been wasted and provides revenue to the owners. The efficacy of the proposed method of economic optimization and stable control operation under different load, PV generation, and central storage conditions is validated by simulation, with results presented to demonstrate its effectiveness. From the simulation results, it is evident that the NLP-based techno-economic optimization gives a maximum revenue of 57 AU$/day to the prosumers. Furthermore, the designed controller is shown to have excellent performance, with accurate power-sharing and voltage regulation capabilities. Future research will be directed towards consideration of the national grid and other renewable energy sources in the system and finding an optimized power-routing scheme for obtaining a successful power solution in the whole country, reducing the pressure on the national grid.

CHAPTER EIGHT

Environmental impact assessment and techno-economic analysis of a hybrid microgrid system

Microgrids (MGs) are playing an important role in the maximum utilization of distributed energy resources. The optimal economic operation and low-carbon electricity generation can enhance MG effectiveness. This chapter presents the results of a solar-photovoltaic (PV)-driven islanded MG's techno-economic optimization analysis and environmental life cycle assessment (LCA) to achieve economically and environmentally superior performance. A net present cost (NPC)-based simulation for optimal sizing of the MG is proposed. A novel life cycle inventory (LCI) is developed to evaluate the impacts of the MG under 21 mid-point indicators and 3 end-point indicators by the ReCiPe 2016 method, and the greenhouse gas (GHG) emissions are evaluated by the IPCC approach. Sensitivity analysis is carried out to verify the effects of three different batteries and five different PV modules on all considered impact indicators. The results reveal that the proposed MG offers a revenue of 29,520 US$/yr by routing excess energy to neighbors after fulfilling the prosumers' demand at an optimal NPC of 364,906 US$. Furthermore, outcomes obtained from the LCA analysis show that among the MG components, batteries have the highest impact on human health (74%) and ecosystems (78%) due to greater GHG emissions (CO_2 48%, CH_4 37%, and N_2O 48%).

8.1. Introduction

In recent years, MGs have received considerable attention worldwide to maximize the utilization of distributed energy resources [225]. MGs are small-scale power systems consisting of distributed generators, loads, an energy storage unit, and a control unit that can be employed in grid-connected or isolated mode for facilitating power supply and/or for maintaining standard service in a distinct locality [226]. Solar-PV-driven islanded MGs offer both profit for the prosumers and much-desired energy for the consumers [200,227]. However, the optimal use of such MGs can be achieved through minimization of the NPC and the levelized cost of energy (COE). Moreover, cleaner electricity generation is pivotal to abating global warming to 2°C by 2030, which is the aim of the 21st Conference of Parties (COP21) of the United Nations Framework Convention on Climate Change (UNFCCC) [228]. Previous literature highlighted that energy-sharing MG frameworks operate cost-effectively at higher demands, but a productive utilization of resources has not been ensured [229,230]. Previous studies also depicted that renewable power plants are responsible for GHG emissions due to fossil fuel consumption in various stages of their lifetimes [15,94], which can be identified by LCA. Therefore, this research aims to optimize cost-economic operation and identify the environmental impact of MGs' elements using a newly created LCI to improve performance for broad application.

The techno-economic analysis of an MG is critical due to the changing nature of its performance with real-time weather data. The modeling of a solar-PV-driven MG in an off-grid islanded situation depends on a few factors, such as the size of the PV modules, the load demand, the size of the converters and inverters, the capacity of storage units, the economics of the elements, energy transmission distances, the amount of excess energy, etc. A number of research groups have used the Hybrid Optimization Model for Electric Renewables (HOMER) for cost–benefit analysis, invented by the US National Renewable Energy Laboratory (NREL) [231,232]. It can handle a wide range of energy sources, such as PV, wind, hydro, fuel cells, boilers, etc., and consumptions such as AC, DC, thermal, hydrogen, etc. Mizani et al. [233] developed a model using HOMER and recognized a best case for the production mix through optimization, which gives lower costs and emissions for the optimal choice of resources. However, they considered a national grid in their model, which is inappropriate for an islanded off-grid community. A standalone MG system is proposed by Thiam et al. considering an island of Senegal, which provides a smaller COE for the community than the national grid [234]. Another standalone MG system that offers a feasible application of renewable resources for a village community is proposed by Lee et al. [194]. Kabir et al. developed an MG system using different renewable resources for electricity production and showed that their system is enough to fulfill the electric demands of an off-grid rural community [206]. Four different cases are modeled and optimized for analyzing the challenges of MG systems using resources from the NREL [235]. A comparative economic assessment of different islanded

MG systems considering diesel, hydro-diesel, and PV-diesel is conducted in [236] and their performance is analyzed. A heuristic algorithm is developed in [237] for modeling an MG system consisting of wind, PV, and battery, and the outcome shows that a proper use of storages can minimize the system's running costs. A case study for a hypothetical locality with a daily load demand of 5000 kWh/day is depicted in [238], but it is not feasible in reality due to unrealistic assumptions. A hybrid source-based system is modeled by Nayar et al. [239] and Anyi et al. [240], in which remote islands of the Maldives and Malaysia are considered, which gives high renewable energy penetration and a solution for remote off-grid application. Givler et al. [229] studied small power systems in Sri Lanka, verified the cost-effectiveness of a PV/diesel-based hybrid MG system, and compared it with a standalone system. Similar have been reported. For example, Himri et al. [241] considered an Algerian village, whereas Nfah et al. [242] and Bekele et al. [243] conducted case studies of Cameroon and Ethiopia, respectively. However, they have not considered the productive use of resources for sharing the excess electricity. Recently, Fernandez et al. proposed a game theory-based energy-sharing model for cost optimization [218], which offers benefits by cutting down the COE through utilizing battery storage at the peak time without using electricity from the national grid. It gives an about 9.17% cost saving in summer. Akter et al. [220] proposed an energy-sharing model using a rule-based approach for energy management, which lacks revenue maximization from an economic perspective. Therefore, NPC-based optimization is a favored method over previously used methods highlighted in [229], [236], [237], [239], [240], [241], [242], [243], [218], [220], and [244] as it optimizes the net present value of the MG system by considering the total annualized cost and the levelized cost. The first main focus of this work is to bridge the research gap by developing an MG system for an islanded locality in Bangladesh, optimizing the system by HOMER, fulfilling the constraints, and routing the excess energy to nearby consumers, not only fulfilling their much-needed energy demands but also offering revenue to the prosumers. The optimization and sensitivity analysis are carried out by systematic priority formation through various optimization techniques.

On the other hand, the life cycle-based environmental impact assessment of an MG is not an easy task as it is required to consider the effects over the lifespan of all elements. Therefore, it is necessary to collect industrial datasets for all MG elements to identify the dangerous releases in their lifetime. The evaluation of various emissions to air, water, and land

and the energy consumption at each life stage of the elements is crucial for LCA analysis. An appropriate strategy is mandatory for assessing and comparing the impacts through LCA. Prior research provides the environmental impacts by each element separately; for example, Liang et al. [17], Notter et al. [245], Hao et al. [246], and Ellingsen et al. [247] examined the effects of lithium-ion batteries, Innocenzi et al. [20] and Meng et al. [21] examined the impacts by the NiMH batteries, and Espinosa et al. [22,23], Goralczyk et al. [164], Gerbinet et al. [25], and Latunussa et al. [24] highlighted the impacts of PV modules. Mizani et al. [233] and Prasai et al. [248] assessed the CO_2 emissions and estimated the reduction of CO_2 release, but they did not consider a systematic LCA approach. Moreover, until now no study has assessed the environmental impacts of MGs based on mid-point and end-point indicators and GHG emissions considering the stages from raw material extraction to end-of-life waste management. The environmental mechanisms for specific impact categories between the inventory data and the category end-points are defined as mid-point indicators, whereas end-point indicators reflect the final effects in a cause–effect chain or of an environmental mechanism [249]. Therefore, the second main focus of this research is to identify the impacts of the proposed MG system using the LCA approach. LCA is a practical method of evaluating the environmental effects of any product, as it identifies the impacts in a broad range of environmental categories, such as resource scarcity, human carcinogenic toxicity, human noncarcinogenic toxicity, ecotoxicity, freshwater eutrophication, terrestrial acidification, ozone formation, global warming, stratospheric ozone depletion, ionizing radiation, water consumption, fine particulate matter formation, land use, etc. [11,120]. LCA analysis deals with the total inputs and outputs, material flows, and emissions at each stage of a product [170]. It also analyzes the lifetime of a product, from the raw material extraction to manufacturing, usage, and end-of-life waste disposal [10,148]. The LCI is developed to assemble the material flows over the lifetimes of the system's elements. The ecoinvent database [172] is used in building the LCI. LCA is accomplished by SimaPro software version 8.5 [250] using the ReCiPe 2016 [151] and Intergovernmental Panel on Climate Change (IPCC) [152] methods. The ReCiPe 2016 method is used to assess the impacts by 18 environmental characterization factors, whereas the IPCC approach is used to identify the GHG emissions (in four categories) of the MG system's elements.

Overall, to optimize costs and reduce the negative impact by the proposed MG, the main contributions of this research can be outlined as follows:
1. A smart MG system is proposed for an islanded remote community, which provides a cost-efficient performance by routing excess electricity to neighboring traditional houses without wasting it off-grid.
2. NPC-based optimization is carried out to maximize profit of prosumers through optimal sizing of elements using real-time physical, operation, and economic inputs in the proposed MG system.
3. A novel LCI is developed that assesses and compares the environmental impacts by each element of the MG using the ReCiPe 2016 and IPCC methods of LCA.
4. Sensitivity analysis is conducted to identify the best cases among various PV modules, such as amorphous silicon (a-Si), copper indium selenide (CIS), multi-Si, ribbon-Si, and single Si, and various community storages, such as lithium-ion (Li-ion), sodium chloride (NaCl), and nickel–metal hydride (NiMH), for lower impact and cost-efficient operation of the MG.

This work is unique in developing an LCI for the LCA analysis of the proposed MG system and minimizing the system cost by optimal sizing of the elements and routing excess energy to neighboring consumers. Given the above purpose, the rest of the chapter is organized as follows. The MG system is introduced in **Section 8.2**. The methods of techno–economic analysis and LCA analysis are discussed in **Section 8.3**. **Section 8.4** highlights the optimal economic operation outcome and the life cycle environmental impact assessment outcome. The sensitivity analysis outcomes considering different cases for PV modules and community storages are presented in **Section 8.5**. Finally, **Section 8.6** presents concluding remarks.

8.2. Microgrid system overview

A typical MG system is an assembly of local energy sources, storages, and loads [196,221]. These MGs play a pivotal role in fulfilling the local load demands of islands and rural villages by power sharing through economic operation [10,251]. In addition, using these MGs the excess energy of the prosumers (with PV facility) can be routed to nearby consumers (without PV facility). It helps to reduce energy wastage to fulfill the much-needed electricity demands of the consumers and to generate profit for the

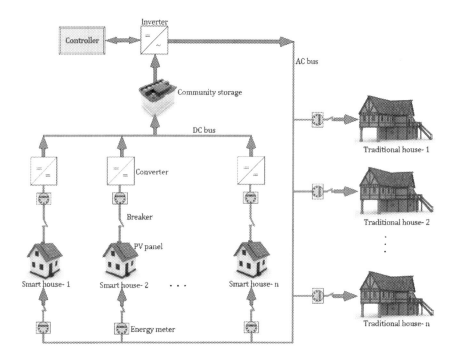

Figure 8.1 The MG framework structure.

prosumers [200,227]. Therefore, MGs are becoming popular from both an economical and a necessity perspective [11,206].

In the proposed MG system, a small array of houses within a short periphery are connected to share energy by an islanded off-grid setup. A few of the houses (smart houses) have a solar-PV electricity production facility, and after fulfilling the prosumers' electricity demands the excess electricity is routed to nearby powerless houses (traditional houses) by this MG system. Fig. 8.1 presents a schematic diagram of the MG system. Overall, in the MG system the PV panels of smart houses, breakers, energy meters, inverters, converters, controller, and community storages are interconnected through cables with a DC bus and an AC bus. The electricity generated by the PV panels is regulated by the DC/DC converters and then directed toward the DC bus. The energy meters record the power flow, a community storage unit stores the generated energy, the inverter converts DC currents to AC, and controllers control the system.

In the model design, the overall electricity generation by the prosumers in normal times is maintained high enough to meet the load demand in an emergency situation in unfavorable weather. Generators are not considered in the model as the system is developed for an islanded situation, with economic constraints for the remote houses. When there exists some renewable energy deficiency in unfavorable weather, the energy would be supplied by their own generators. On the other hand, when surplus energy is generated in some hours, prosumers make some profit by routing this [244,252,253]. A central storage unit is used instead of decentralized ones for convenient sharing of energy among the smart houses, for routing the overall extra power to the traditional houses, and for lowering the NPC [254].

The NPC-based optimization model provides the optimal PV module number and community storage size for the considered load demands of the prosumers and calculates the amount of excess energy for routing to the traditional houses, as described in **Section 8.3.1**. A novel LCI is developed to assess the lifetime environmental impacts of the considered MG system using LCA, as discussed in **Section 8.3.2**.

8.3. Methods

Two different methods are used in this research: one for optimal economic operation and another for environmental impact assessment of the MG system. These methods are discussed in the following subsections.

8.3.1 Optimal economic operation

The chosen remote village for this research is Kutubdia, a small island in the Bay of Bengal in the Cox's Bazar district of Bangladesh. The MG provides off-grid electricity to the inhabitants of this village, as it has no access to the national grid.

The total NPC and the levelized COE are dependent on the total yearly expense of the MG system. The overall yearly expense of the MG is the sum of its elements' expenses minus miscellaneous expenses. The following equation is used to calculate the NPC of the MG [231]:

$$NPC = \frac{C_{Total}}{CRF_{(\eta,n)}}, \qquad (8.1)$$

where C_{Total} is the overall yearly expenses of the MG, η is the interest rate per year, n is the year number, and $CRF_{(\eta,n)}$ is the capital recovery. $CRF_{(\eta,n)}$

and the COE are calculated as follows:

$$CRF_{(\eta,n)} = \frac{\eta(1+\eta)^n}{(1+\eta)^{n-1}}, \quad (8.2)$$

$$COE = \frac{C_{Total}}{E_P + E_C}, \quad (8.3)$$

where E_P and E_C are the annual load demands of prosumers and consumers, respectively, met by the MG system. The optimization is carried out for a minimal NPC of the MG system using HOMER following the method described in [231,232].

The installed PV capacity is defined as follows [255]:

$$P_{PV} = V_{PV} \times I_{PV}, \quad (8.4)$$

$$V_{PV} = \frac{mkT}{q} \ln(1 + \frac{I_{SC}}{I_0}), \quad (8.5)$$

$$I_{PV} = I_{SC} - I_0(e^{\frac{qV_{PV}}{mkT}} - 1), \quad (8.6)$$

where V_{PV} is the output voltage of each PV cell, I_{PV} is the PV current of each cell, m is the ideality factor, k is the Boltzmann constant, T is the PV cell temperature, I_{SC} is the short-circuit current, I_0 is the saturation current, and q is the charge of an electron.

The community storage is managed by its storage capacity using the following [255]:

$$C_x = \frac{V \times D \times SoC}{Y_{converter} \times Y_{storage}}, \quad (8.7)$$

where C_x is the storage capacity, x is the discharge time in hours, V is the voltage in offload conditions, D is the number of zero charging days, $Y_{converter}$ and $Y_{storage}$ are converter and storage yield values, respectively, and SoC is the state of charge.

The key assumptions for the simulation of the MG system using HOMER are as follows.

8.3.1.1 Model parameters

The proposed MG system has a total of 12 houses, of which 4 are considered as smart houses with a PV facility, while the others are traditional

Table 8.1 Simulation parameters.

Parameter	Value
Solar scaled average	4.5 kWh/m^2/day
Nominal capacity of community storage	1820 kWh
Each PV panel capacity	60 kW
Inverter and converter capacity	72 kW
Prosumer load demand	370 kWh/day
Each PV panel cost	6956 US$
Community storage cost	1006 US$
Inverter/converter cost	1671 US$
Community storage replacement cost	533 US$
Inverter/converter replacement cost	472 US$
Each PV panel capacity factor	18%
Interest rate	0.5%
Project lifetime	25 years

houses that depend on smart houses for electricity, as there is no connection to the national grid. The considered model parameters for NPC-based optimization are presented in Table 8.1. The consumers' loads are not considered as the main aim is to fulfill prosumers' demands first and rout the remaining electricity to consumers after optimization, which will generate revenue for the prosumers. The solar radiation profile of Cox's Bazar, Bangladesh, is used for this work, which is collected from the NASA Surface Meteorology website [256]. The average solar radiation is found to be 4.5 kWh/m^2/day. All expenses associated with capital, operation, maintenance, replacement, and fuel and miscellaneous expenses are included in the NPC. All expenses of the MG system are considered in constant dollars [231,232].

8.3.1.2 Simulation

HOMER simulation is conducted to identify the lifetime cost feasibility and the optimal operation strategy for the MG system. It runs simulations on an hourly basis and it considers the sustainable operation capacity of the grid.

8.3.1.3 Optimization

HOMER provides the optimal sizing of the PV panels and battery strings through optimization maintaining the constraints. It considers minimum NPC for the system and gives an optimal configuration after optimization.

8.3.1.4 Sensitivity analysis

Sensitivity analysis is conducted to identify the influences of changing various parameters of the system. In this research, various lifetimes of batteries, such as 5, 10, 15, and 20 years, and various solar scale capacities, such as 4.5, 5, and 5.58, are considered for sensitivity analysis. Many optimization outcomes are achieved for these assumptions and are used to identify the best case.

8.3.2 Life cycle environmental impact assessment method

LCA is a systematic approach to environmental impact evaluation to identify and categorize the effects caused by a product or process throughout its entire lifetime [111,113]. This approach consists of four basic steps: (i) goal and scope definition, (ii) LCI, (iii) life cycle impact estimation, and (iv) life cycle impact interpretation. For LCA analysis following these steps, International Standardization Organization (ISO) standards 14040:2006 and 14044:2006 are followed [118,119]. In the below subsections, the LCA steps are briefly described to provide the LCA methodology followed in this study.

8.3.2.1 Goal and scope definition

The first LCA step is goal and scope definition, where the objective is defined and the LCA system boundaries are established. The goal of this LCA is to identify and compare the negative environmental impacts of the MG system. The scope of this LCA analysis is cradle-to-grave [117,120] for mid-point and end-point environmental impact indicators for the system. Therefore, the comprehensive LCA considers the lifetime of the system including raw material extraction, key parts manufacturing, transportation, MG system installation, and end-of-life waste disposal. The functional unit [111,113] of this LCA is chosen as 1 kWh of electricity supplied by the MG system.

8.3.2.2 Life cycle inventory

The second step is the development of the LCI, where all inputs (material and energy) and outputs (emissions) at each stage of the elements' lifetimes are added. The formation of the LCA boundary of the MG system as shown in (Fig. 8.2) is the unique contribution of this work. The boundary is modeled with mandatory equipment during stages such as raw material extraction from mines, transportation of these materials, production of MG

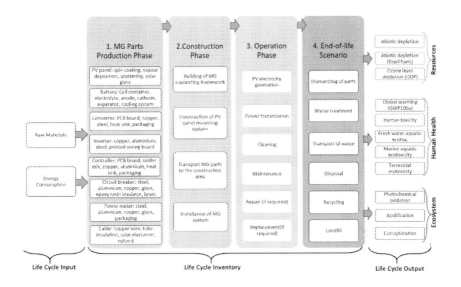

Figure 8.2 The system boundary of the MG framework for LCA analysis.

parts, transportation of the parts to the MG location, MG construction, MG operation, and end-of-life waste disposal.

The intakes and releases at each stage are schematically shown in Fig. 8.3. Energy and material intake take place in the raw material extraction, MG element production, and MG installation stages, whereas only energy intake happens in the transportation and waste management stages. Solid materials and gases are emitted during several stages, such as waste management, MG installation, MG element production, and raw material extraction. Energy output is only found in the MG operation stage. Fig. 8.4 shows the stage-by-stage energy and material flows for the considered MG system. The ecoinvent database [121,122,257] is used to collect the life cycle inputs and outputs of the MG elements because it contains global industrial and commercial datasets for the manufacture, transportation, waste management, etc., of different elements [101,121,122]. From the ecoinvent database, base unit processes are chosen depending on item specifications. Table 8.2 gives the data source for the MG elements. The considered PV panel and battery (community storage) types are global CIS and Li-ion, respectively. Moreover, the global unit processes are considered for other elements such as the inverter, converter, cable, breaker, and energy meter. An assembly of these unit processes of the elements is formed for the desired MG system, which is finally used to evaluate the individual effects

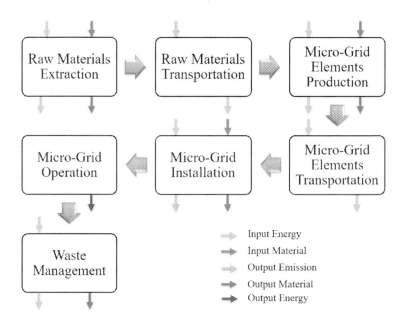

Figure 8.3 The stage-wise material, energy, and emission flow.

Table 8.2 Data collection for LCA of the MG framework.

Unit process	Process source
PV panel	Photovoltaic panel, CIS, at plant/GLO U/I U/AusSD U
Battery	Battery, Li-ion, rechargeable, prismatic {GLO}/ market for / Alloc Def, U
Energy meter	Electric meter, unspecified {GLO}/ production / Conseq, U
Breaker	Switch, toggle, type, at plant/GLO U/AusSD U
Cable	Cable, unspecified, {GLO}/ market for / Conseq, U
Converter	Converter, 250 W, for electric system {GLO}/ production / Conseq, U
Inverter	Inverter, 250 W, at plant/GLO U/I U/AusSD U

by every process element. The energy losses and heat releases during the power transmission and distribution stages are not considered in this LCA due to a lack of datasets.

8.3.2.3 Life cycle impact estimation

In the third LCA step, LCA is carried out based on the ISO 14040:2006 standard following the ReCiPe 2016, IPCC, and Eco-points methods.

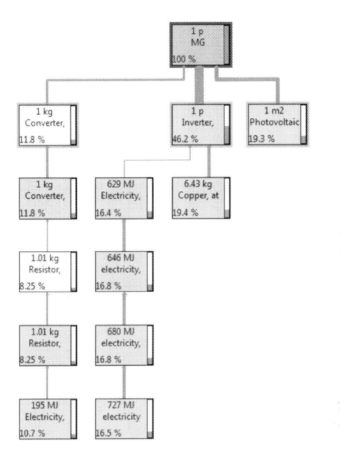

Figure 8.4 The material flow of the MG framework.

SimaPro software version 8.5 [250] is used to identify the effects after developing the LCA system boundary because it is universally for LCA [138]. The ReCiPe 2016 method [151] is used for assessing the mid-point indicators because this approach combines the scientific rigor of the CML2001 and the Eco-indicator-99 approaches and considers 18 mid-point categories (8 more than Eco-indicator-99), which is highest among all methods. The 18 mid-point impact indicators assessed by the ReCiPe 2016 method are resource scarcity (fossil and mineral), human carcinogenic toxicity, human noncarcinogenic toxicity, ecotoxicity (marine, terrestrial, and freshwater), freshwater eutrophication, terrestrial acidification, ozone formation (human health and terrestrial ecosystems), global warming (human health, terrestrial ecosystems, and freshwater ecosystems), stratospheric

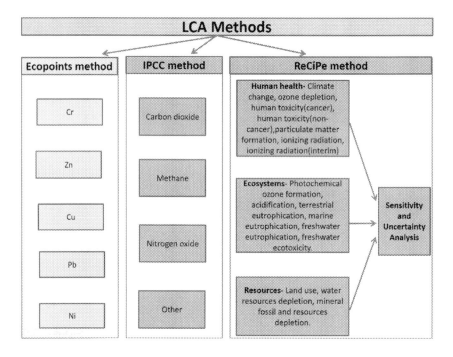

Figure 8.5 The LCA methods used in this analysis.

ozone depletion, ionizing radiation, water consumption (human health, terrestrial ecosystems, and aquatic ecosystems), fine particulate matter formation, and land use. The ReCiPe 2016 method is used to assess the end-point indicators because this method considers three aggregated categories: human health, ecosystems, and resources (Fig. 8.5).

Moreover, the IPCC approach is used to assess the emission of GHGs such as carbon dioxide, methane, nitrogen oxide, etc. (Fig. 8.5), following a 100-year time frame [152]. This method provides three advantages in evaluating the release of GHGs: (i) it ensures optimal utilization of available datasets in a comprehensive way, (ii) it provides accuracy in estimation, and (iii) it provides information for policy developers on climate change. It does not deal with emission of carbon monoxide and radiative substances in the environment.

Furthermore, the Eco-points method is utilized to quantify the release of harmful metal particles such as lead (Pb), copper (Cu), cadmium (Cd), chromium (Cr), zinc (Zn), and nickel (Ni) from each element of the MG (Fig. 8.5). This approach uses three steps to assess the impacts: classification, normalization, and weighting. Classification includes dataset collection and cumulative summation for each element for each impact indicator from the whole life cycle of the system. In the normalization step, the magnitude of environmental degradation is determined by taking the ratio of the impact indicator and a reference value. In the weighting phase of the Eco-points method, the impact indicator value is multiplied with the weighting factor to obtain the single-score value for comparative analysis.

8.3.2.4 Life cycle impact interpretation

In the last LCA step, life cycle interpretation, impact outcomes are analyzed and interpreted to identify the most significant substances for each of the mid-point and end-point environmental effect indicators over the lifetime of the MG system. These findings are correlated with the sustainability factors of the systems to abate the impacts on human health, ecosystems, and resources.

Finally, sensitivity analysis is carried out by the ReCiPe 2016 method, considering various PV panels (single Si, multi-Si, ribbon-Si, CIS, and a-Si) and community storages (NiMH, NaCl, and Li–ion) to check the changes in effects. This helps to identify the best option with regard to the environment impact aspects.

8.4. Results and discussion

8.4.1 Economic operation outcome

The obtained optimal operation is presented in Table 8.3. It is found that having three PV panels along with the community storage and converters is the optimal case with the lowest NPC of the MG system. The optimal NPC and COE rates are 364,906 US$ and 0.139 US$, whereas, the MG system with four PV panels is the worst case with an NPC of 442,574 US$ and a COE of 0.169 US$. It is also found that about 79,783 kWh/yr (22.5%) of excess electricity is produced, so a profit of 29,520 US$/yr (considering 0.37 US$/kWh) is generated through routing to nearby consumers. The yearly scenario of the total renewable power output and the excess electrical production is depicted in Fig. 8.6. The outcome shows that there is a significant amount of excess energy in each month of the year, which can be routed to more traditional houses to generate revenue for the prosumers.

The results are compared with existing studies of standalone MG systems, revealing that a range of design parameters such as PV mod-

Table 8.3 The NPC-based optimization result of the MG framework.

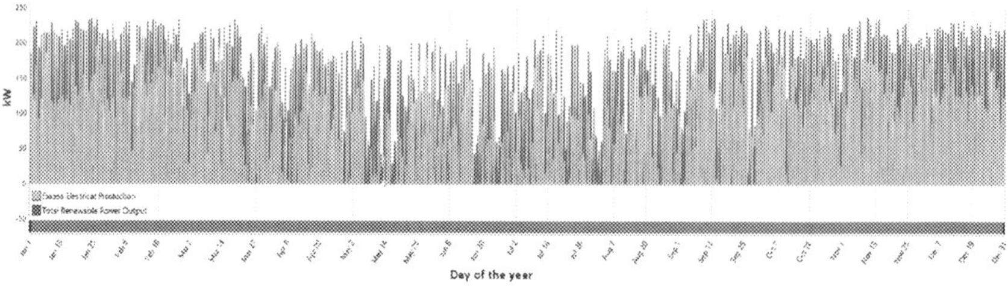

Figure 8.6 The annual excess power rate of the MG framework.

ule size, community storage capacity, solar irradiation rate, converter/inverter capacity, storage strings number, etc., are responsible for different COEs ranging between 0.1 US$ and 2 US$ [194,234,241–243]. The obtained COE in [234] is 0.18 US$/kWh for Senegal, in [194] it is 0.49 US$/kWh for China, in [242] it is 0.34 US$/kWh for Cameroon, and in [243] it is 0.94 US$/kWh for Ethiopia, while in this study the COE is 0.13 US$/kWh for Bangladesh, which is very small. On the other hand, the obtained NPC in [241] is 2,982,825 US$ for Algeria, in [242] it is 376,856 US$ for Cameroon, and in [243] it is 464,600 US$ for Ethiopia, whereas in this study it is 364,906 US$ for Bangladesh.

8.4.2 Life cycle environmental impact assessment results
8.4.2.1 Environmental impacts of the microgrid

The mid-point impacts of MG parts obtained by a cradle-to-grave analysis using the ReCiPe 2016 method are presented in Fig. 8.7. It is found that the community storage has the highest impact for most of the indicators, such as water consumption (82.68%), terrestrial ecotoxicity (70.05%), ozone formation (68.87%), and global warming (68.51%), whereas the inverter has a very low impact on indicators such as ionizing radiation (0.47%), freshwater ecotoxicity (0.80%), stratospheric ozone depletion (1.15%), and land use (1.25%). The cable is most harmful for mineral resource scarcity (54.40%) and least harmful for water consumption (0.01%). The breaker has the highest impacts on freshwater eutrophication (14.77%) and the second highest on ozone formation (6.97%). The energy meter is significantly af-

Environmental impact assessment and techno-economic analysis of a hybrid MG system 195

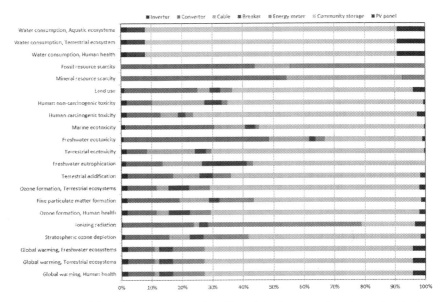

Figure 8.7 The life cycle environmental profiles of the framework as determined using the ReCiPe 2016 method.

fects the impact categories of ionizing radiation (50.37%) and fossil resource scarcity (44.35%). The converter has the lowest effects on water consumption (0.04%) and is optimal for mineral resource scarcity (54.40%). The PV panel impact is about 9.32% for water consumption (human health, terrestrial ecosystem, and aquatic ecosystem). The PV panel impact is about 4.23% for global warming (human health, terrestrial ecosystem, and aquatic ecosystem). Overall, the PV modules had smaller effects on most of the mid-point categories such as mineral and fossil resource scarcity (0.001%), freshwater eutrophication (0.09%), marine ecotoxicity (0.40%), and human noncarcinogenic toxicity (0.65%).

The comparative end-point impacts of the MG elements obtained by the ReCiPe 2016 method are highlighted in Fig. 8.8, which shows that the converter, energy meter, and cables impact resources mostly with rates of 45%, 39%, and 15%, respectively. Community storage affects the ecosystems and human health greatly with rates of 79% and 74%, respectively, due to their large size and dangerous chemicals. The PV panels had the highest effects on ecosystems due to significant amounts of fossil fuel consumption, mostly in the raw material extraction and processing phases, and high end-of-life pollution. Therefore, future research should be conducted to utilize

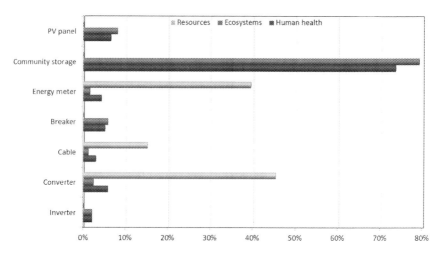

Figure 8.8 End-point damage assessment of the framework using the ReCiPe 2016 method.

renewable resources in all stages of the MG components' lifetime from raw material extraction to end-of-life recycling and waste disposal.

The most impactful substances for the end-point indicators of each MG element are depicted in Table 8.4, which shows that carbon dioxide is released from the community storage, greatly affecting human health (3.18E−05 DALY/kWh). Zinc from the converter is most harmful for human health (1.31E−04 DALY/kWh) and copper from cables affects mostly human health (4.09E−05 DALY/kWh). Hard coal is responsible for the impacts of the PV panels on resources (9.54E−06 kg/kWh) and natural gas is responsible for the impact of cables (6.39E−07 kg/kWh). Overall, the dangerous substances that most affect human health, resources, and ecosystems are copper, zinc, particulate matter, silver, carbon dioxide, and sulfur dioxide.

8.4.2.2 Greenhouse gas emissions by the microgrid

The life cycle GHG emissions from the MG components, obtained by the IPCC method, are compared in Fig. 8.9. It is found that carbon dioxide is mostly emitted for the community storage and PV panels, with rates of about 48% and 34%, respectively. Moreover, the release of methane and nitrous oxide is also high for the community storage, with rates of 38% and 48%, respectively. The converter and energy meter release most to the land during end-of-life recycling with rates of 43% and 42.5%, respectively. Overall, community storage is highly dangerous due to the maximum re-

Table 8.4 The key hazardous substances of the MG elements that mostly affect the end-point environmental indicators.

Impact Category	Community storage	Converter	Cable	Switch	Meter	Inverter	PV panel
Human health (DALY/kWh)	Carbon dioxide (3.18E-05)	Zinc (1.31E-04)	Copper (4.09E-05)	Chromium (9.63E-05)	Particulates (5.14E-05)	Sulfur dioxide (2.32E-04)	Sulfur oxides (3.58E-04)
Ecosystems (species.yr/kWh)	Phosphate (7.94E-07)	Ammonia (8.16E-08)	Zinc (6.25E-08)	Nitrogen (7.84E-07)	Sulfur hexafluoride (2.53E-08)	Phosphate (2.74E-07)	Carbon dioxide (4.64E-06)
Resources (kg/kWh)	Zirconium (4.97E-06)	Silver (1.39E-04)	Natural gas (6.39E-07)	Xylene (1.94E-08)	Copper (2.16E-08)	Aluminum (4.05E-07)	Hard coal (9.54E-06)

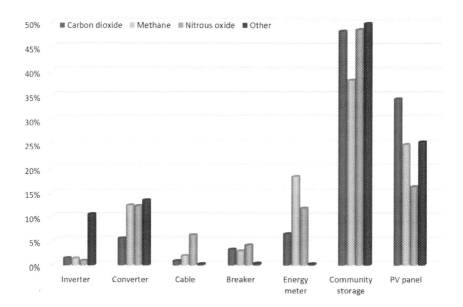

Figure 8.9 GHG emission as determined using the IPCC method.

lease of GHGs for all three categories of carbon dioxide, methane, and nitrous oxide due to the chemicals used. Therefore, researchers should pay considerable attention to enhancing the environmental profiles of community storage units.

8.4.2.3 Metal particles release by the microgrid

The release of metal particles as determined by the Eco-points 97 method is depicted in Fig. 8.10. The solar panel contributes most to metal particle emissions in the categories of Zn, Cu, Cd, and Ni. In contrast, the inverter is the least impactful based on metal particle release to the environment. Pb and Cr are mostly released by the community storage with rates of 38% and 71%, respectively. However, the converter, cable, breaker, and energy meter show a medium risk to the environment with respect to metal particle release.

8.5. Sensitivity analysis

Several sensitivity analyses have been carried out to test the technoeconomic operation and the environmental profiles of the proposed MG

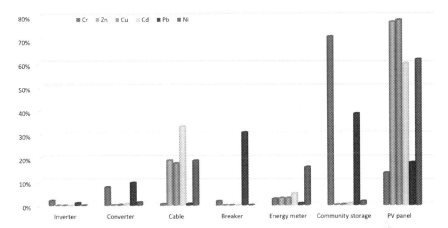

Figure 8.10 The metal-based emissions quantification outcome using the Eco-points 97 method.

system for various battery lifetimes, solar scales, PV modules, and community storages, which helped to identify the best case.

8.5.1 Effects for various community storage lifetimes and solar scale rates

The effect of changing the community storage's lifetime and the solar scales is shown in Table 8.5, which reveals that the lowest NPC (364,906 US$) and COE (0.139 US$) values are found for the case of a 20-year life and 5.58 kWh/m^2/day solar irradiation, which is the optimized case among the 12 options. On the other hand, a community storage with a 5-year life and 4.50 kWh/m^2/day solar irradiation provides the maximum NPC of 462,351 US$ and the highest COE of 0.177 US$. Moreover, the same COE of 0.171 US$ is obtained for the three cases with various combinations of community storage lifetimes and solar scales. Overall, the increased community storage lifetime and solar scale provide better sensitivity outcomes for the MGs, with lower COE and NPC values.

8.5.2 Impacts for various types of PV modules

The sensitivity analysis outcome using five different PV modules such as single Si, multi-Si, ribbon-Si, a-Si, and CIS for the considered MG system is depicted in Table 8.6. It is found that multi-Si is notably responsible for water consumption (85.98 mPt), whereas CIS has the lowest impact on water consumption (21.47 mPt). The a-Si PV modules are highly account-

Table 8.5 Sensitivity analysis outcomes for various battery lifetimes and solar-scaled factors in NPC-based optimization of the MG.

able for ionizing radiation (37.87 mPt) and the single-Si PV modules are mostly responsible for water consumption (84.73 mPt) and global warming (45.14 mPt). Overall, the CIS-based PV modules give a better environmental performance for most of the impact indicators. Therefore, investors should use CIS solar modules in building MG systems due to their superior environmental profile, to reduce the environmental impacts.

8.5.3 Impacts of various types of community storages

The sensitivity analysis outcome using various community storages for the MG system, shown in Table 8.7, highlights that an NiMH-based system has stronger effects for most of the categories, such as stratospheric ozone depletion (32.07 mPt), water consumption (21.11 mPt), ionizing radiation (18.05 mPt), terrestrial acidification (15.12 mPt), fine particulate matter formation (14.25 mPt), and global warming (3.65 mPt). The NaCl community storage-based MG system has a maximum impact for three categories: mineral resource scarcity (25.33 mPt), fossil resource scarcity (9.12 mPt), and freshwater ecotoxicity (5.91 mPt). Overall, sensitivity analysis of the Li–ion community storage-based MG system exhibits the best environmental performance. Therefore, prosumers should use Li–ion type community storage units in constructing MG systems. The key implication of the sensitivity analysis outcome lies in smart grids, in which shareholders should use environment-friendly community storage units to avoid environmental danger.

8.6. Conclusion

In this chapter, an NPC-based optimization analysis and an LCA-based environmental impact assessment of a solar-PV-driven off-grid MG framework is presented. To ensure the validity of this research, we (i) developed an off-grid MG system, (ii) optimized this based on NPC minimiza-

Table 8.6 Sensitivity analysis outcome for various PV modules of the MG.

Impact category	Single-Si [mPt]	Multi-Si [mPt]	Ribbon-Si [mPt]	a-Si [mPt]	CIS [mPt]
Global warming, human health	45.14	37.75	34.51	19.99	33.91
Global warming, terrestrial ecosystems	45.14	37.75	34.51	19.99	33.91
Global warming, freshwater ecosystems	45.14	37.75	34.50	19.99	33.91
Stratospheric ozone depletion	41.46	37.58	32.61	11.18	15.97
Ionizing radiation	14.30	14.89	11.79	37.85	4.16
Ozone formation, human health	37.92	31.19	28.88	12.55	29.19
Fine particulate matter formation	23.14	17.14	16.22	9.38	18.68
Ozone formation, terrestrial ecosystems	38.01	31.51	29.24	12.35	28.65
Terrestrial acidification	28.54	21.37	20.52	11.11	23.61
Freshwater eutrophication	3.73	3.44	3.35	0.86	1.91
Terrestrial ecotoxicity	33.81	33.27	33.15	2.13	4.05
Freshwater ecotoxicity	14.65	10.96	9.32	6.74	5.87
Marine ecotoxicity	18.52	18.24	17.99	2.23	2.97
Human carcinogenic toxicity	26.55	21.54	16.66	20.14	16.75
Human noncarcinogenic toxicity	22.19	21.87	21.58	3.57	5.66
Land use	47.57	42.95	38.52	16.18	31.62
Water consumption, human health	84.73	85.98	78.53	35.09	21.47
Water consumption, terrestrial ecosystem	84.73	85.98	78.53	35.09	21.47
Water consumption, aquatic ecosystems	84.73	85.98	78.53	35.09	21.47

tion, (iii) analyzed the life cycle material flow, (iv) built an LCI, (v) assessed environmental profiles by multiple methods, and (vi) conducted sensitivity analyses that examine the optimal design and environmental performance of the MG. The well-known HOMER Pro and SimaPro software programs and the renowned ecoinvent global database were used for the cost

Table 8.7 Sensitivity analysis outcome for various batteries of the MG.

Impact category	NiMH [mPt]	NaCl [mPt]	Li-ion [mPt]
Global warming, human health	3.65	1.19	2.13
Global warming, terrestrial ecosystems	3.65	1.19	2.13
Global warming, freshwater ecosystems	3.65	1.19	2.13
Stratospheric ozone depletion	32.07	2.48	1.15
Ionizing radiation	18.05	13.42	0.47
Ozone formation, human health	3.20	1.39	2.10
Fine particulate matter formation	14.25	11.26	1.95
Ozone formation, terrestrial ecosystems	3.17	1.39	2.08
Terrestrial acidification	15.12	11.64	2.16
Freshwater eutrophication	1.98	1.47	1.62
Terrestrial ecotoxicity	2.60	1.89	1.85
Freshwater ecotoxicity	1.11	5.91	0.80
Marine ecotoxicity	2.18	1.90	1.46
Human carcinogenic toxicity	2.17	3.31	1.99
Human noncarcinogenic toxicity	0.64	1.70	1.89
Land use	1.28	1.01	1.25
Mineral resource scarcity	0	25.33	0
Fossil resource scarcity	0	9.12	0
Water consumption, human health	21.11	0.01	2.02
Water consumption, terrestrial ecosystem	21.11	0.01	2.02
Water consumption, aquatic ecosystems	21.11	0.01	2.02

optimization and impact assessment. This research is unique in developing an LCI and assessing the impacts of an MG by multiple methods such as ReCiPe 2016 for mid-point and end-point effects analysis and IPCC for GHG emission estimation. Results reveal that the NPC-based techno-economic optimization offers a profit of 29,520 US$/yr to the prosumers at an optimal NPC of 364,906 US$ and a levelized COE of 0.139 US$. Furthermore, the LCA outcome shows that the battery is the most impactful element of the MG for most of the mid-point impact indicators, such as global warming (68.51%), land use (59.45%), ecotoxicity (32.12%), eutrophication (56.79%), and acidification (62.25%). The sensitivity analysis outcome highlights that increases in the lifetime of the community storage unit and the solar scale provide minimal NPC and that CIS-PV modules and Li-ion batteries are environmentally superior to others for MGs. Connection to a national grid, incorporation of other renewable sources in the MG framework, and replacement of dangerous elements with environment-friendly materials is the future direction of this research in the quest for broader application and cleaner operation.

CHAPTER NINE

Future directions towards green and sustainable energy

This book has extended our understanding of the environmental effects of each of the elements of renewable energy systems such as solar-photovoltaic (PV), hydro, wind, and biomass plants for cleaner production of electricity and the conceptual framework of islanded microgrids (MGs) for optimal economic operation. This research has compared the impacts of renewable plants located in the US and revealed the best option considering the smallest effect for all impact indicators, including global warming, ozone formation, ecotoxicity, water consumption, acidification, eutrophication, ionizing radiation, carcinogenic radiation, ozone depletion, and land use. Moreover, fossil fuel-based power consumption and greenhouse gas (GHG) emissions have been evaluated. This work has identified an optimal smart power-routing MG framework, which routes any excess energy to neighbor loads, utilizing energy that would otherwise have been wasted and generating revenue for the prosumers. The general concluding remarks of this book and future research directions are highlighted below.

9.1. Book summary and concluding remarks

9.1.1 Summary

This book has presented the challenges of cleaner energy production and economic operation of renewable energy technologies. In order to facilitate this, novel life cycle inventories (LCIs) are developed to assess the effects of solar-PV, solar-thermal, wind, biomass, and hydropower plants. Additionally, a unique islanded MG framework is proposed with verified case studies for economic optimization and stable control operation under different load, PV generation, and central storage conditions.

Although state-of-the-art resources and tools have been employed recently in evaluating the environmental impacts of renewable power plants throughout their lifespan, there is still a research gap in identifying the key processes which need to be the focus of attention. The review findings reveal that the assessment indicators "resources" and "ecosystems" are key factors that are mostly lacking in the literature, even though they are crucial for people in locations nearby plants. **Chapter 2** analyzes and summarizes the existing literature to identify the research gaps in the field of sustainable renewable electricity generation. It also provides important insights into the research gaps where the focus should be in the development of low-carbon electricity production technologies for a sustainable future.

A systematic life cycle assessment (LCA) is performed to assess the environmental effects of solar-PV and solar-thermal frameworks, which is presented in **Chapter 3**. The assessment is carried out by the International Life Cycle Data System (ILCD), cumulative energy demand (CED), Eco-indicator 99, and Intergovernmental Panel on Climate Change (IPCC) methods using the newly developed LCI considering life cycle inputs and outputs. The findings are compared to make better-informed choices. Uncertainty and sensitivity analyses have also been conducted for both frameworks, which identified the superior storage and solar collector type for most of the considered impact categories.

The quantification of the ecological effects of hydropower plants located in alpine and nonalpine areas of Europe is presented in **Chapter 4**. A novel comprehensive LCI is developed to assess the environmental hazards by both plant categories in terms of ecosystems, climate change, resources, and human health. Several systematic LCA methods such as ReCiPe 2016, Impact 2002+, and Eco-points 97 are used to identify impacts based on various impact indicators such as global warming, ozone formation, ecotoxicity, water consumption, acidification, eutrophication, ionizing radiation, carcinogenic radiation, ozone depletion, and land use. This is the first approach in assessing the effects from both installations considering both the mid-point and the end-point approaches.

A comparative analysis of the environmental effects of solar-PV, biomass, and pumped storage hydropower plants in the US is presented in **Chapter 5**. The analysis is carried out by the Eco-indicator 99, Tool for the Reduction and Assessment of Chemical and other Environmental Impacts (TRACI), Raw Material Flows, CED, and IPCC methods. The emissions of hazardous metals and GHGs from the considered plants are estimated over their lifetime. Additionally, the fossil fuel-based energy consumption during the manufacturing of plant elements, the installation and operation of the plants, and waste management is evaluated for the plants.

A comparative analysis of the environmental impacts of four different renewable power plants, namely solar, wind, biomass, and hydropower plants, through a systematic LCA analysis by the CML, Eco-indicator 99, Eco-points 97, Raw Material Flows, and IPCC methods using the ecoinvent database and SimaPro software is presented in **Chapter 6**. The results reveal that the hydropower plant is the least damaging to the environment as it emits lower levels of GHGs, CO_2, NO_x, and SO_x than others. Moreover, the total environmental impacts for all of the end-point indicators, like human health, ecosystem quality, and resources, are lower for hydropower plants than for others. The biomass power plant has a high environmental impact, whereas solar and wind power plants cause medium damage to the environment. The outcomes from this research will guide governments and investors in taking appropriate decisions by installing the best renewable power plants considering the environmental aspects and available resources.

A new power-sharing approach by a PV/central storage system (CSS) hybrid source-based remote area MG framework is proposed in **Chapter 7**. This research work is novel in developing a nonlinear programming-based optimization model for day-ahead scheduling of available power routing using the proposed smart power-routing framework, which routes any excess energy to neighbor loads, utilizing energy that would otherwise have been wasted and generating revenue to the owners. The efficacy of the proposed method of economic optimization and stable control operation under different load, PV generation, and central storage conditions is validated by simulation, and the results are presented to demonstrate its effectiveness.

A net present cost (NPC)-based optimization analysis and an LCA-based environmental impact assessment of a solar-PV-driven off-grid MG framework is presented in **Chapter 8**. The well-known HOMER Pro and SimaPro software and the renowned ecoinvent global database are used for the cost optimization and impact assessment. This research is unique in developing an LCI and assessing the impacts of an MG by multiple methods such as ReCiPe 2016 for mid-point and end-point effects analysis and IPCC for GHG emission estimation.

9.1.2 Conclusions

The concluding remarks of this research work are as follows:
- Solar-PV systems are less impactful for human health, climate change, and ecosystems than solar-thermal systems.
- Copper indium selenide (CIS)-solar collectors perform better than amorphous Si, multi-Si, ribbon-Si, and single Si for solar-thermal and solar-PV systems.
- Li-ion batteries perform environmentally better than NaCl and NiMH for the solar technologies.

- The global warming effects of hydropower plants in nonalpine regions (3.92×10^{-4} kg CO_2-eq./MJ) are higher than those of plants in alpine regions (2.97×10^{-5} kg CO_2-eq./MJ).
- Alpine region-based hydropower plants release one-tenth of the amount of methane released by nonalpine region-based plants.
- Solar-PV plants are the most environmentally sustainable in the US as they emit less carbon dioxide (4.03×10^{-6} kg CO_2-eq./kWh), methane (1.62×10^{-7} kg CO_2-eq./kWh), and nitrogen oxide (1.42×10^{-7} kg CO_2-eq./kWh) than biomass and pumped storage hydropower plants.
- Pumped storage hydropower plants in the US are a threat to the environment based on their environmental impact assessment as they consume fossil fuels or nuclear energy for pumping water to the storage, whereas biomass power plants cause medium damage to the environment except for the categories of smog and ozone layer depletion.
- The life cycle GHG emissions from solar-PV plants are lower (6.42×10^{-6} kg CO_2-eq./kWh) than those of wind (1.93×10^{-4} kg CO_2-eq./kWh) and hydropower plants (2.71×10^{-5} kg CO_2-eq./kWh).
- Fossil fuel consumption rates of wind plants are higher than those of solar-PV and hydropower plants, as determined by the CED approach.
- Hydropower plants have lower metal emissions than wind and solar-PV plants in their whole lifetime, as determined using the Eco-points 97 method.
- Solar-PV plants in Switzerland perform environmentally better than hydro and wind plants for 9 out of 10 impact indicators.
- The amount of total daily excess energy for routing which would have been wasted if not shared is found to be approximately 109.57 kWh. Therefore, the total revenue of a typical day is about 57.32 AU$ (20,635.2 AU$/yr) with the proposed framework.
- The designed controller is shown to have excellent performance, with accurate power-sharing and voltage regulation capabilities.
- The NPC-based techno-economic optimization offers a profit of 29,520 US$/yr to the prosumers at an optimal NPC of 364,906 US$ and a levelized cost of energy of 0.139 US$.
- The LCA outcome shows that the battery is the MG element with the highest environmental impact for most of the mid-point impact indicators, such as global warming (68.51%), land use (59.45%), ecotoxicity (32.12%), eutrophication (56.79%), and acidification (62.25%).
- The sensitivity analysis outcome highlights that an increased lifetime of the community storage unit and solar scale provides minimal NPC and

that CIS-PV modules and Li-ion batteries are environmentally superior to others for an MG.

9.2. Future research directions

Future research directions are provided below.
- The replacement of the power plant elements that are responsible for hazardous emissions without affecting efficiency and robustness of the system must be studied in future.
- Ways to reduce the consumption of fossil fuels during the manufacturing of elements, installation and operation of the system, and waste management must be investigated in the future.
- The energy payback period of renewable systems must be estimated in future.
- Sensitivity analysis must be conducted to replace all impactful elements of considered power plants by sustainable alternatives.
- Future research should be concentrated on investigating the effects of and discovering solutions to environmental hazards of installed power plants.
- Connection to a national grid and incorporation of other renewable sources are necessary for MG frameworks. Moreover, dangerous elements should be replaced with environment-friendly alternatives to achieve cleaner operation.
- Future research should include the development of a new framework by considering a combination of renewable and nonrenewable energy sources and finding an optimized power-routing scheme, reducing the pressure on the national grid.

APPENDIX A

List of acronyms

ABC	Three-phase components
AC	Alternating current
a-Si	Amorphous silicon
BESS	Battery energy storage system
CED	Cumulative energy demand
CIS	Copper indium selenide
CML	Center of Environmental Science of Leiden University of Sweden
COD	Dissolved organic carbon
COE	Cost of energy
CRF	Capital recovery factor
CSS	Central storage system
DALY	Disability-adjusted life years
DC	Direct current
DQO	Direct, quadrature, and zero components
EPBT	Energy payback time
GHG	Greenhouse gas
HH	Human health
HHV	Higher heating values
HOMER	Hybrid Optimization Model for Electric Renewables
HTF	Heat transfer fluid
ILCD	International Life Cycle Data System
IPCC	Intergovernmental Panel on Climate Change
ISO	International Standardization Organization
LCA	Life cycle assessment
LCI	Life cycle inventory
LCL	Inductor, capacitor, and inductor
LHV	Lower heating values
MG	Microgrid
MGC	Microgrid controller
NPC	Net present cost
NREL	National Renewable Energy Laboratory
PM	Particulate matter
PV	Photovoltaic
PVT	Photovoltaic thermal
RES	Renewable energy system
RET	Renewable energy technology
RMF	Raw Material Flows
SoC	State of charge
Species.yr	Species-years
TRACI	Tool for the Reduction and Assessment of Chemical and other Environmental Impacts
UNFCCC	United Nations Framework Convention on Climate Change

Green Energy
https://doi.org/10.1016/B978-0-32-385953-0.00016-1

Copyright © 2023 Elsevier Inc.
All rights reserved.

211

APPENDIX B

List of symbols

P	Active power
Q	Reactive power
ω	Angular frequency
f	Frequency
$*$	Reference values for corresponding variables
D_P, D_Q	Droop coefficients of active and reactive power controller
$P^s_{t,h}$	Power supplied to house h during t
$P^d_{t,h}$	Power demand of house h during t
t	Index of time periods
h	Index of houses
j	Index of PV systems
ΔT	Duration of time periods (hours)
$P^g_{t,j}$	Power generation of a PV system
H_S	Set of smart houses (with PV facility)
H_T	Set of traditional houses (without PV facility)
G	Set of PV generation units
λ_t	Per-unit cost for traded energy during period t
SoC_t	State of charge of the CSS during period t
E_{cap}	Maximum storage capacity of the CSS
η_{ch}	Charging efficiency of the CSS
η_{dis}	Discharging efficiency of the CSS
C_{Total}	Overall yearly expenses of the MG
η	Interest rate per year
n	Year number
$CRF_{(\eta,n)}$	Capital recovery for year n
E_P	Annual load demands of prosumers
E_C	Annual load demands of consumers
V_{PV}	Output voltage of each PV cell
I_{PV}	PV current of each cell
I_{PV}	PV power of each cell
m	Ideality factor
k	Boltzmann's constant
T	PV cell temperature
I_{SC}	Short-circuit current
I_0	Saturation current
q	Charge of an electron
C_x	Storage capacity
x	Discharge time in 1 hour
V	Voltage in offload condition
D	Zero charging days
$Y_{converter}$	Converter yield value
$Y_{storage}$	Storage yield value

kg CO_2-eq.	Kilograms of carbon dioxide equivalent
kg CFC-11-eq.	Kilograms of CFC-11 equivalents
kg Cl-eq.	Kilograms of chlorine equivalents
kg Cd-eq.	Kilograms of cadmium equivalents
tkm	Tonne-kilometers
kg N-eq.	Kilograms of nitrogen equivalent
kg O_3-eq.	Kilograms of ozone equivalent
MJ LHD	Megajoules of lower heating values
MJ HHD	Megajoules of higher heating values
kg 1,4-DB-eq.	Kilograms of 1,4-dichlorobenzene equivalents
kg C_2H_4-eq.	Kilograms of ethylene equivalents
kg PO_4-eq.	Kilograms of phosphate equivalents
kg SO_2-eq.	Kilograms of sulfur dioxide equivalents
kg NO_2-eq.	Kilograms of nitrogen dioxide equivalents
CTUh	Comparative toxic unit for humans
kg PM 2.5-eq.	Kilograms of PM2.5 equivalents
kg CH_4-eq.	Kilograms of methane equivalents
kg N-eq.	Kilograms of nitrogen equivalents
PDF*m^2yr	The yearly potentially disappeared fraction per unit area
PAF*m^2yr	The yearly potentially affected fraction per unit area
CTUe	Comparative toxic unit for ecosystems
kg N_2O	Kilograms of nitrous oxide equivalents
MJ primary	Total life cycle primary energy use

References

[1] M. Curran, Life-cycle assessment, in: S.E. Jorgensen, B.D. Fath (Eds.), Encyclopedia of Ecology, Academic Press, Oxford, 2008, pp. 2168–2174.

[2] M.A.P. Mahmud, N. Huda, S.H. Farjana, C. Lang, A strategic impact assessment of hydropower plants in alpine and non-alpine areas of Europe, Applied Energy 250 (2019) 198–214.

[3] P.W.M. in the Alps, Situation report on hydropower generation in the Alpine region focusing on small hydropower, in: A Platform within the Alpine Convention, vol. 1, 2010, pp. 1–52.

[4] F. Manzano-Agugliaro, M. Taher, A. Zapata-Sierra, A. Juaidi, F.G. Montoya, An overview of research and energy evolution for small hydropower in Europe, Renewable & Sustainable Energy Reviews 75 (2017) 476–489.

[5] G. Rebitzer, T. Ekvall, R. Frischknecht, D. Hunkeler, G. Norris, T. Rydberg, W.-P. Schmidt, S. Suh, B. Weidema, D. Pennington, Life cycle assessment: Part 1: Framework, goal and scope definition, inventory analysis, and applications, Environment International 30 (5) (2004) 701–720.

[6] U.S. energy information administration, monthly energy review, appendix d.1, and tables 1.1 and 10.1, Preliminary data for 2017.

[7] S.H. Farjana, N. Huda, M.P. Mahmud, R. Saidur, A review on the impact of mining and mineral processing industries through life cycle assessment, Journal of Cleaner Production 231 (2019) 1200–1217.

[8] U.S. energy information administration, Electric Power Monthly, Chapters 1 and 3, with data for August 2018.

[9] R. Yuan, J.F. Rodrigues, A. Tukker, P. Behrens, The impact of the expansion in non-fossil electricity infrastructure on China's carbon emissions, Applied Energy 228 (2018) 1994–2008.

[10] M.A.P. Mahmud, N. Huda, S.H. Farjana, C. Lang, Environmental impacts of solar-photovoltaic and solar-thermal systems with life-cycle assessment, Energies 11 (9) (2018) 2346.

[11] M.A.P. Mahmud, N. Huda, S.H. Farjana, C. Lang, Environmental sustainability assessment of hydropower plant in Europe using life cycle assessment, IOP Conference Series: Materials Science and Engineering 351 (1) (2018) 012006.

[12] N. Agrawal, M. Ahiduzzaman, A. Kumar, The development of an integrated model for the assessment of water and GHG footprints for the power generation sector, Applied Energy 216 (2018) 558–575.

[13] Y. Mu, W. Cai, S. Evans, C. Wang, D. Roland-Holst, Employment impacts of renewable energy policies in China: A decomposition analysis based on a CGE modeling framework, Applied Energy 210 (2018) 256–267.

[14] M.V. Barros, R. Salvador, C.M. Piekarski, A.C. de Francisco, F.M.C.S. Freire, Life cycle assessment of electricity generation: a review of the characteristics of existing literature, The International Journal of Life Cycle Assessment (2019) 4–19.

[15] Z. Li, H. Du, Y. Xiao, J. Guo, Carbon footprints of two large hydro-projects in China: Life-cycle assessment according to ISO/TS 14067, Renewable Energy 114 (2017) 534–546, http://www.sciencedirect.com/science/article/pii/S0960148117307012.

[16] L. Lelek, J. Kulczycka, A. Lewandowska, J. Zarebska, Life cycle assessment of energy generation in Poland, The International Journal of Life Cycle Assessment 21 (1) (2016) 1–14.

[17] Y. Liang, J. Su, B. Xi, Y. Yu, D. Ji, Y. Sun, C. Cui, J. Zhu, Life cycle assessment of lithium-ion batteries for greenhouse gas emissions, Resources, Conservation and Recycling 117 (2017) 285–293.

[18] D. Akinyele, J. Belikov, Y. Levron, Battery storage technologies for electrical applications: Impact in stand-alone photovoltaic systems, Energies 10 (11) (2017) 1760.

[19] M.A.P. Mahmud, N. Huda, S.H. Farjana, C. Lang, Comparative life cycle environmental impact analysis of lithium-ion (LiIo) and nickel-metal hydride (NiMH) batteries, Batteries 5 (1) (2019) 22–28.

[20] V. Innocenzi, N.M. Ippolito, I.D. Michelis, M. Prisciandaro, F. Medici, F. Veglio, A review of the processes and lab-scale techniques for the treatment of spent rechargeable NiMH batteries, Journal of Power Sources 362 (2017) 202–218.

[21] T. Meng, K.-H. Young, D.F. Wong, J. Nei, Ionic liquid-based non-aqueous electrolytes for nickel/metal hydride batteries, Batteries 3 (1) (2017).

[22] N. Espinosa, M. Hosel, M. Jorgensen, F.C. Krebs, Large scale deployment of polymer solar cells on land, on sea and in the air, Energy & Environmental Science 7 (2014) 855–866, https://doi.org/10.1039/C3EE43212B.

[23] N. Espinosa, Y.-S. Zimmermann, G.A. dos Reis Benatto, M. Lenz, F.C. Krebs, Outdoor fate and environmental impact of polymer solar cells through leaching and emission to rainwater and soil, Energy & Environmental Science 9 (2016) 1674–1680, https://doi.org/10.1039/C6EE00578K.

[24] C.E. Latunussa, F. Ardente, G.A. Blengini, L. Mancini, Life cycle assessment of an innovative recycling process for crystalline silicon photovoltaic panels, Solar Energy Materials and Solar Cells 156 (2016) 101–111, http://www.sciencedirect.com/science/article/pii/S0927024816001227.

[25] S. Gerbinet, S. Belboom, A. Léonard, Life cycle analysis (LCA) of photovoltaic panels: A review, Renewable & Sustainable Energy Reviews 38 (2014) 747–753, http://www.sciencedirect.com/science/article/pii/S136403211400495X.

[26] J. Hou, W. Zhang, P. Wang, Z. Dou, L. Gao, D. Styles, Greenhouse gas mitigation of rural household biogas systems in China: A life cycle assessment, Energies 10 (2) (2017) 239.

[27] B. Atilgan, A. Azapagic, Life cycle environmental impacts of electricity from fossil fuels in Turkey, in: Bridges for a more sustainable future: Joining Environmental Management for Sustainable Universities (EMSU) and the European Roundtable for Sustainable Consumption and Production (ERSCP) conferences, Journal of Cleaner Production 106 (2015) 555–564, http://www.sciencedirect.com/science/article/pii/S095965261400763X.

[28] M.P. Mahmud, S.H. Farjana, Wind power technology schemes as renewable energy in Bangladesh, International Journal of Engineering and Advanced Technology (IJEAT) 1 (2012) 315–319, http://www.ijeat.org/download/volume-1-issue-5/.

[29] D.J. Ward, O.R. Inderwildi, Global and local impacts of UK renewable energy policy, Energy & Environmental Science 6 (2013) 18–24, https://doi.org/10.1039/C2EE22342B.

[30] P.D.D. Das, P.R. Srinivasan, P.D.A. Sharfuddin, Fossil fuel consumption, carbon emissions and temperature variation in India, Energy & Environment 22 (6) (2011) 695–709, https://doi.org/10.1260/0958-305X.22.6.695.

[31] L.M. Rubio, J.P.B. Filho, J.R. Henriquez, Performance of a PV/T solar collector in a tropical monsoon climate city in Brazil, IEEE Latin America Transactions 16 (1) (2018) 140–147.

[32] E. Santoyo-Castelazo, H. Gujba, A. Azapagic, Life cycle assessment of electricity generation in Mexico, Energy 36 (3) (2011) 1488–1499, http://www.sciencedirect.com/science/article/pii/S0360544211000193.

[33] M.Z. Jacobson, M.A. Delucchi, G. Bazouin, Z.A.F. Bauer, C.C. Heavey, E. Fisher, S.B. Morris, D.J.Y. Piekutowski, T.A. Vencill, T.W. Yeskoo, 100% clean and renewable wind, water, and sunlight (WWS) all-sector energy roadmaps for the 50 United States, Energy & Environmental Science 8 (2015) 2093–2117, https://doi.org/10.1039/C5EE01283J.

[34] S. Srinivasan, N. Kholod, V. Chaturvedi, P.P. Ghosh, R. Mathur, L. Clarke, M. Evans, M. Hejazi, A. Kanudia, P.N. Koti, B. Liu, K.S. Parikh, M.S. Ali, K. Sharma, Water for electricity in India: A multi-model study of future challenges and linkages to climate change mitigation, Applied Energy 210 (2018) 673–684.

[35] L. Gaudard, F. Avanzi, C.D. Michele, Seasonal aspects of the energy-water nexus: The case of a run-of-the-river hydropower plant, Applied Energy 210 (2018) 604–612.

[36] A.B. Hidrovo, J. Uche, A. Martínez-Gracia, Accounting for GHG net reservoir emissions of hydropower in Ecuador, Renewable Energy 112 (2017) 209–221.

[37] L. Scherer, S. Pfister, Global water footprint assessment of hydropower, Renewable Energy 99 (2016) 711–720.

[38] R. Turconi, A. Boldrin, T. Astrup, Life cycle assessment (LCA) of electricity generation technologies: Overview, comparability and limitations, Renewable & Sustainable Energy Reviews 28 (2013) 555–565, http://www.sciencedirect.com/science/article/pii/S1364032113005534.

[39] T. Huang, J. Wang, Research on charging and discharging control strategy of electric vehicles and its economic benefit in microgrid, in: 2016 IEEE International Conference on Power and Renewable Energy (ICPRE), vol. 2, 2016, pp. 518–522.

[40] J. Chipindula, V. Botlaguduru, H. Du, R. Kommalapati, Z. Huque, Life cycle environmental impact of onshore and offshore wind farms in Texas, Sustainability 10 (2018) 2020–2022.

[41] L. Xu, M. Pang, L. Zhang, W.-R. Poganietz, S.D. Marathe, Life cycle assessment of onshore wind power systems in China, Resources, Conservation and Recycling 132 (2018) 361–368, http://www.sciencedirect.com/science/article/pii/S0921344917301684.

[42] A. Schreiber, J. Marx, P. Zapp, Comparative life cycle assessment of electricity generation by different wind turbine types, Journal of Cleaner Production 233 (2019) 561–572.

[43] R. Fang, Life cycle cost assessment of wind power–hydrogen coupled integrated energy system, International Journal of Hydrogen Energy (2019).

[44] E. Beagle, E. Belmont, Comparative life cycle assessment of biomass utilization for electricity generation in the European Union and the United States, Energy Policy 128 (2019) 267–275.

[45] P.O. Loução, J.P. Ribau, A.F. Ferreira, Life cycle and decision analysis of electricity production from biomass – Portugal case study, Renewable & Sustainable Energy Reviews 108 (2019) 452–480.

[46] J.M. Maier, T. Sowlati, J. Salazar, Life cycle assessment of forest-based biomass for bioenergy: A case study in British Columbia, Canada, Resources, Conservation and Recycling 146 (2019) 598–609.

[47] S. You, H. Tong, J. Armin-Hoiland, Y.W. Tong, C.-H. Wang, Techno-economic and greenhouse gas savings assessment of decentralized biomass gasification for electrifying the rural areas of Indonesia, Applied Energy 208 (2017) 495–510.

[48] W. Luo, Y.S. Khoo, A. Kumar, J.S.C. Low, Y. Li, Y.S. Tan, Y. Wang, A.G. Aberle, S. Ramakrishna, A comparative life-cycle assessment of photovoltaic electricity generation in Singapore by multicrystalline silicon technologies, Solar Energy Materials and Solar Cells 174 (2018) 157–162, http://www.sciencedirect.com/science/article/pii/S0927024817304877.

[49] M.Z. Akber, M.J. Thaheem, H. Arshad, Life cycle sustainability assessment of electricity generation in Pakistan: Policy regime for a sustainable energy mix, Energy Policy 111 (2017) 111–126.

[50] T.A. Quek, W.A. Ee, W. Chen, T.A. Ng, Environmental impacts of transitioning to renewable electricity for Singapore and the surrounding region: A life cycle assessment, Journal of Cleaner Production 214 (2019) 1–11.

[51] C. Perilhon, D. Alkadee, G. Descombes, S. Lacour, Life cycle assessment applied to electricity generation from renewable biomass, in: Terragreen 2012: Clean Energy Solutions for Sustainable Environment (CESSE), Energy Procedia 18 (2012) 165–176, http://www.sciencedirect.com/science/article/pii/S1876610212007989.

[52] F. Ardente, G. Beccali, M. Cellura, V.L. Brano, Life cycle assessment of a solar thermal collector, Renewable Energy 30 (7) (2005) 1031–1054.

[53] D. Nugent, B.K. Sovacool, Assessing the lifecycle greenhouse gas emissions from solar PV and wind energy: A critical meta-survey, Energy Policy 65 (2014) 229–244, http://www.sciencedirect.com/science/article/pii/S0301421513010719.

[54] J. Kabayo, P. Marques, R. Garcia, F. Freire, Life-cycle sustainability assessment of key electricity generation systems in Portugal, Energy 176 (2019) 131–142.

[55] A. Roinioti, C. Koroneos, Integrated life cycle sustainability assessment of the Greek interconnected electricity system, Sustainable Energy Technologies and Assessments 32 (2019) 29–46.

[56] O. Siddiqui, I. Dincer, Comparative assessment of the environmental impacts of nuclear, wind and hydro-electric power plants in Ontario: A life cycle assessment, Journal of Cleaner Production 164 (2017) 848–860, http://www.sciencedirect.com/science/article/pii/S0959652617314063.

[57] F. de Miranda Ribeiro, G.A. da Silva, Life-cycle inventory for hydroelectric generation: a Brazilian case study, Journal of Cleaner Production 18 (1) (2010) 44–54, http://www.sciencedirect.com/science/article/pii/S0959652609002777.

[58] Varun, I. Bhat, R. Prakash, LCA of renewable energy for electricity generation systems—a review, Renewable & Sustainable Energy Reviews 13 (5) (2009) 1067–1073, http://www.sciencedirect.com/science/article/pii/S1364032108001093.

[59] M.A. Curran, Life cycle assessment: a review of the methodology and its application to sustainability, in: Energy and environmental engineering / Reaction engineering and catalysis, Current Opinion in Chemical Engineering 2 (3) (2013) 273–277, http://www.sciencedirect.com/science/article/pii/S2211339813000221.

[60] R. Dones, U. Gantner, Greenhouse gas emissions from hydropower full energy chain in Switzerland, in: Assessment of Greenhouse Gas Emission from the Full Energy Chain for Hydropower, Nuclear Power and Other Energy Sources-IAEA Advisory Group Meeting, Hydro-Quebec, Montreal, Canada, 1996, pp. 12–14.

[61] Y. Tripanagnostopoulos, M. Souliotis, R. Battisti, A. Corrado, Energy, cost and LCA results of PV and hybrid PV/T solar systems, Progress in Photovoltaics: Research and Applications 13 (2005) 235–250.

[62] R. Kannan, K. Leong, R. Osman, H. Ho, C. Tso, Life cycle assessment study of solar PV systems: An example of a 2.7kwp distributed solar PV system in Singapore, Solar Energy 80 (5) (2006) 555–563.

[63] V.M. Fthenakis, H.C. Kim, E. Alsema, Emissions from photovoltaic life cycles, Environmental Science & Technology 42 (6) (2008) 2168–2174.

[64] T.-T. Chow, J. Ji, Environmental life-cycle analysis of hybrid solar photovoltaic/thermal systems for use in Hong Kong, International Journal of Photoenergy 2012 (2012) 101968–101977.

[65] Y. Fu, X. Liu, Z. Yuan, Life-cycle assessment of multi-crystalline photovoltaic (PV) systems in China, Journal of Cleaner Production 86 (2015) 180–190.

[66] M. Yu, A. Halog, Solar photovoltaic development in Australia—a life cycle sustainability assessment study, Sustainability 7 (2015) 1213–1247.
[67] P. Wu, X. Ma, J. Ji, Y. Ma, Review on life cycle assessment of energy payback of solar photovoltaic systems and a case study, Energy Procedia 105 (2017) 68–74.
[68] J. Eskew, M. Ratledge, M. Wallace, S.H. Gheewala, P. Rakkwamsuk, An environmental life cycle assessment of rooftop solar in Bangkok, Thailand, Renewable Energy 123 (2018) 781–792.
[69] G. Martinopoulos, Life cycle assessment of solar energy conversion systems in energetic retrofitted buildings, Journal of Building Engineering 20 (2018) 256–263.
[70] N.A. Ludin, N.I. Mustafa, M.M. Hanafiah, M.A. Ibrahim, M.A.M. Teridi, S. Sepeai, A. Zaharim, K. Sopian, Prospects of life cycle assessment of renewable energy from solar photovoltaic technologies: A review, Renewable & Sustainable Energy Reviews 96 (2018) 11–18.
[71] F. Liu, T. Lv, Assessment of geographical distribution of photovoltaic generation in China for a low carbon electricity transition, Journal of Cleaner Production 212 (2019) 655–665.
[72] O.B. Mousa, S. Kara, R.A. Taylor, Comparative energy and greenhouse gas assessment of industrial rooftop-integrated PV and solar thermal collectors, Applied Energy 241 (2019) 113–123.
[73] H. Wang, E. Oguz, B. Jeong, P. Zhou, Life cycle and economic assessment of a solar panel array applied to a short route ferry, Journal of Cleaner Production 219 (2019) 471–484.
[74] A. Pascale, T. Urmee, A. Moore, Life cycle assessment of a community hydroelectric power system in rural Thailand, Renewable Energy 36 (11) (2011) 2799–2808.
[75] M. Pang, L. Zhang, C. Wang, G. Liu, Environmental life cycle assessment of a small hydropower plant in China, The International Journal of Life Cycle Assessment 20 (6) (2015) 796–806.
[76] A. Kadiyala, R. Kommalapati, Z. Huque, Evaluation of the life cycle greenhouse gas emissions from hydroelectricity generation systems, Sustainability 8 (6) (2016) 539.
[77] G. Mao, S. Wang, Q. Teng, J. Zuo, X. Tan, H. Wang, Z. Liu, The sustainable future of hydropower: A critical analysis of cooling units via the theory of inventive problem solving and life cycle assessment methods, Journal of Cleaner Production 142 (2017) 2446–2453.
[78] T. Ueda, E. Roberts, A. Norton, D. Styles, A. Williams, H. Ramos, J. Gallagher, A life cycle assessment of the construction phase of eleven micro-hydropower installations in the UK, Journal of Cleaner Production 218 (2019) 1–9.
[79] Y.-F. Huang, X.-J. Gan, P.-T. Chiueh, Life cycle assessment and net energy analysis of offshore wind power systems, Renewable Energy 102 (2017) 98–106.
[80] M.N. Perez-Camacho, R. Curry, T. Cromie, Life cycle environmental impacts of biogas production and utilisation substituting for grid electricity, natural gas grid and transport fuels, Waste Management 95 (2019) 90–101.
[81] S. Alanya-Rosenbaum, R.D. Bergman, Life-cycle impact and exergy based resource use assessment of torrefied and non-torrefied briquette use for heat and electricity generation, Journal of Cleaner Production 233 (2019) 918–931.
[82] A. Ozawa, Y. Kudoh, N. Kitagawa, R. Muramatsu, Life cycle CO_2 emissions from power generation using hydrogen energy carriers, International Journal of Hydrogen Energy 44 (2019) 11219–11232.
[83] M. Ozturk, I. Dincer, Life cycle assessment of hydrogen-based electricity generation in place of conventional fuels for residential buildings, International Journal of Hydrogen Energy (2019).
[84] R. Contreras-Lisperguer, E. Batuecas, C. Mayo, R. Díaz, F. Pérez, C. Springer, Sustainability assessment of electricity cogeneration from sugarcane bagasse in Jamaica, Journal of Cleaner Production 200 (2018) 390–401.

[85] E. Benetto, E.-C. Popovici, P. Rousseaux, J. Blondin, Life cycle assessment of fossil CO_2 emissions reduction scenarios in coal-biomass based electricity production, Energy Conversion and Management 45 (18) (2004) 3053–3074.

[86] J. Hanafi, A. Riman, Life cycle assessment of a mini hydro power plant in Indonesia: A case study in Karai river, in: The 22nd CIRP Conference on Life Cycle Engineering, Procedia CIRP 29 (2015) 444–449.

[87] M.T.B. Geller, A.A.d.M. Meneses, Life cycle assessment of a small hydropower plant in the Brazilian Amazon, Journal of Sustainable Development of Energy, Water and Environment Systems 4 (4) (2016) 379–391.

[88] S. Righi, F. Baioli, A. Dal Pozzo, A. Tugnoli, Integrating life cycle inventory and process design techniques for the early estimate of energy and material consumption data, Energies 11 (4) (2018) 970.

[89] B. Atilgan, A. Azapagic, Assessing the environmental sustainability of electricity generation in Turkey on a life cycle basis, Energies 9 (1) (2016) 31.

[90] R. Kommalapati, A. Kadiyala, M.T. Shahriar, Z. Huque, Review of the life cycle greenhouse gas emissions from different photovoltaic and concentrating solar power electricity generation systems, Energies 10 (3) (2017) 350.

[91] A.T. Dale, A.F. Pereira de Lucena, J. Marriott, B.S.M.C. Borba, R. Schaeffer, M.M. Bilec, Modeling future life-cycle greenhouse gas emissions and environmental impacts of electricity supplies in Brazil, Energies 6 (7) (2013) 3182–3208.

[92] D.J. Murphy, M. Carbajales-Dale, D. Moeller, Comparing apples to apples: Why the net energy analysis community needs to adopt the life-cycle analysis framework, Energies 9 (11) (2016) 917.

[93] M.P. Mahmud, S.H. Farjana, Design and construction of refrigerant charge level detecting device in HVAC/R system with microcontroller, International Journal of Engineering and Advanced Technology (IJEAT) 1 (2012) 309–314, http://www.ijeat.org/download/volume-1-issue-5/.

[94] N.Y. Amponsah, M. Troldborg, B. Kington, I. Aalders, R.L. Hough, Greenhouse gas emissions from renewable energy sources: A review of lifecycle considerations, Renewable & Sustainable Energy Reviews 39 (2014) 461–475, http://www.sciencedirect.com/science/article/pii/S1364032114005395.

[95] M.P. Mahmud, J. Lee, G. Kim, H. Lim, K.-B. Choi, Improving the surface charge density of a contact-separation-based triboelectric nanogenerator by modifying the surface morphology, in: Micro/Nano Devices and Systems 2015, Microelectronic Engineering 159 (2016) 102–107, http://www.sciencedirect.com/science/article/pii/S0167931716301058.

[96] S. Lizin, S. Van Passel, E. De Schepper, W. Maes, L. Lutsen, J. Manca, D. Vanderzande, Life cycle analyses of organic photovoltaics: a review, Energy & Environmental Science 6 (2013) 3136–3149, https://doi.org/10.1039/C3EE42653J.

[97] I. Capellan-Perez, I. Arto, J.M. Polanco-Martinez, M. Gonzalez-Eguino, M.B. Neumann, Likelihood of climate change pathways under uncertainty on fossil fuel resource availability, Energy & Environmental Science 9 (2016) 2482–2496, https://doi.org/10.1039/C6EE01008C.

[98] E. El-Bialy, S. Shalaby, A. Kabeel, A. Fathy, Cost analysis for several solar desalination systems, Desalination 384 (2016) 12–30.

[99] A.C. Tamboli, D.C. Bobela, A. Kanevce, T. Remo, K. Alberi, M. Woodhouse, Low-cost CdTe/silicon tandem solar cells, IEEE Journal of Photovoltaics 7 (6) (2017) 1767–1772.

[100] S.H. Farjana, N. Huda, M.A.P. Mahmud, Life-cycle environmental impact assessment of mineral industries, IOP Conference Series: Materials Science and Engineering 351 (2018) 012016.

[101] I. Boustead, General principles for life cycle assessment databases, in: Environmental Life Cycle Assessment and its Applications, Journal of Cleaner Production 1 (3) (1993) 167–172, http://www.sciencedirect.com/science/article/pii/095965269390008Y.
[102] J. Gong, S.B. Darling, F. You, Perovskite photovoltaics: life-cycle assessment of energy and environmental impacts, Energy & Environmental Science 8 (2015) 1953–1968, https://doi.org/10.1039/C5EE00615E.
[103] A. Stoppato, Life cycle assessment of photovoltaic electricity generation, in: 19th International Conference on Efficiency, Cost, Optimization, Simulation and Environmental Impact of Energy Systems, Energy 33 (2) (2008) 224–232, http://www.sciencedirect.com/science/article/pii/S0360544207002137.
[104] M.A.P. Mahmud, N. Huda, S.H. Farjana, M. Asadnia, C. Lang, Recent advances in nanogenerator-driven self-powered implantable biomedical devices, Advanced Energy Materials 8 (2) (2018) 172–210, https://onlinelibrary.wiley.com/doi/abs/10.1002/aenm.201701210.
[105] N. Arnaoutakis, M. Souliotis, S. Papaefthimiou, Comparative experimental life cycle assessment of two commercial solar thermal devices for domestic applications, Renewable Energy 111 (2017) 187–200.
[106] N.A. Masruroh, B. Li, J. Klemes, Life cycle analysis of a solar thermal system with thermochemical storage process, Renewable Energy 31 (4) (2006) 537–548.
[107] G. Hou, H. Sun, Z. Jiang, Z. Pan, Y. Wang, X. Zhang, Y. Zhao, Q. Yao, Life cycle assessment of grid-connected photovoltaic power generation from crystalline silicon solar modules in China, Applied Energy 164 (2016) 882–890, http://www.sciencedirect.com/science/article/pii/S0306261915014646.
[108] E. Leccisi, M. Raugei, V. Fthenakis, The energy and environmental performance of ground-mounted photovoltaic systems—a timely update, Energies 9 (8) (2016) 622.
[109] A. Sherwani, J. Usmani, Varun, Life cycle assessment of solar PV based electricity generation systems: A review, Renewable & Sustainable Energy Reviews 14 (1) (2010) 540–544, http://www.sciencedirect.com/science/article/pii/S1364032109001907.
[110] T. Chow, A review on photovoltaic/thermal hybrid solar technology, Applied Energy 87 (2) (2010) 365–379.
[111] R. Heijungs, J.B. Guineev, An Overview of the Life Cycle Assessment Method – Past, Present, and Future, Wiley-Blackwell, 2012, ch. 2, pp. 15–41, https://onlinelibrary.wiley.com/doi/abs/10.1002/9781118528372.ch2.
[112] S.H. Farjana, N. Huda, M.A.P. Mahmud, C. Lang, Towards sustainable TiO_2 production: An investigation of environmental impacts of ilmenite and rutile processing routes in Australia, Journal of Cleaner Production 196 (2018) 1016–1025.
[113] A. Lewandowska, A. Matuszak-Flejszman, K. Joachimiak, A. Ciroth, Environmental life cycle assessment (LCA) as a tool for identification and assessment of environmental aspects in environmental management systems (EMS), The International Journal of Life Cycle Assessment 16 (3) (Mar 2011) 247–257, https://doi.org/10.1007/s11367-011-0252-3.
[114] G.A. Keoleian, The application of life cycle assessment to design, in: Environmental Life Cycle Assessment and its Applications, Journal of Cleaner Production 1 (3) (1993) 143–149, http://www.sciencedirect.com/science/article/pii/095965269390004U.
[115] C. Pesso, Life cycle methods and applications: issues and perspectives, in: Environmental Life Cycle Assessment and its Applications, Journal of Cleaner Production 1 (3) (1993) 139–142, http://www.sciencedirect.com/science/article/pii/095965269390003T.
[116] J.A. de Larderel, Environmental life cycle assessment and its applications, in: Environmental Life Cycle Assessment and its Applications, Journal of Cleaner

Production 1 (3) (1993) 130, http://www.sciencedirect.com/science/article/pii/095965269390001R.
[117] H.A.U. de Haes, Methodology and LCA applications, in: Environmental Life Cycle Assessment and its Applications, Journal of Cleaner Production 1 (3) (1993) 205–206, http://www.sciencedirect.com/science/article/pii/095965269390020C.
[118] J. Pryshlakivsky, C. Searcy, Fifteen years of ISO 14040: a review, Journal of Cleaner Production 57 (2013) 115–123, http://www.sciencedirect.com/science/article/pii/S0959652613003764.
[119] M. Finkbeiner, A. Inaba, R. Tan, K. Christiansen, H.-J. Klüppel, The new international standards for life cycle assessment: ISO 14040 and ISO 14044, The International Journal of Life Cycle Assessment 11 (2) (Mar 2006) 80–85, https://doi.org/10.1065/lca2006.02.002.
[120] P. Stavropoulos, C. Giannoulis, A. Papacharalampopoulos, P. Foteinopoulos, G. Chryssolouris, Life cycle analysis: Comparison between different methods and optimization challenges, in: Research and Innovation in Manufacturing: Key Enabling Technologies for the Factories of the Future - Proceedings of the 48th CIRP Conference on Manufacturing Systems, Procedia CIRP 41 (2016) 626–631, http://www.sciencedirect.com/science/article/pii/S2212827115011270.
[121] J. Pascual-González, G. Guillén-Gosálbez, J.M. Mateo-Sanz, L. Jiménez-Esteller, Statistical analysis of the ecoinvent database to uncover relationships between life cycle impact assessment metrics, Journal of Cleaner Production 112 (2016) 359–368, http://www.sciencedirect.com/science/article/pii/S0959652615007088.
[122] R. Frischknecht, G. Rebitzer, The ecoinvent database system: a comprehensive web-based LCA database, in: Life Cycle Assessment, Journal of Cleaner Production 13 (13) (2005) 1337–1343, http://www.sciencedirect.com/science/article/pii/S0959652605001253.
[123] O. Jolliet, M. Margni, R. Charles, S. Humbert, J. Payet, G. Rebitzer, R. Rosenbaum, Impact 2002+: A new life cycle impact assessment methodology, The International Journal of Life Cycle Assessment 8 (6) (2003) 324.
[124] L. Mancini, L. Benini, S. Sala, Resource footprint of Europe: Complementarity of material flow analysis and life cycle assessment for policy support, Environmental Science & Policy 54 (2015) 367–376.
[125] M. Rohrlich, M. Mistry, P.N. Martens, S. Buntenbach, M. Ruhrberg, M. Dienhart, S. Briem, R. Quinkertz, Z. Alkan, K. Kugeler, A method to calculate the cumulative energy demand (CED) of lignite extraction, The International Journal of Life Cycle Assessment 5 (6) (Nov 2000) 369–373, https://doi.org/10.1007/BF02978675.
[126] J.A. Cherni, R. Olalde Font, L. Serrano, F. Henao, A. Urbina, Systematic assessment of carbon emissions from renewable energy access to improve rural livelihoods, Energies 9 (12) (2016) 1086.
[127] E. Neri, D. Cespi, L. Setti, E. Gombi, E. Bernardi, I. Vassura, F. Passarini, Biomass residues to renewable energy: A life cycle perspective applied at a local scale, Energies 9 (11) (2016) 922.
[128] L. Sheng-Qiang, M. Xian-Qiang, G. Yu-Bing, X. You-Kai, Life cycle assessment, estimation and comparison of greenhouse gas mitigation potential of new energy power generation in China, Advances in Climate Change Research 3 (3) (2012) 147–153.
[129] D. Garcia-Gusano, D. Iribarren, D. Garrain, Prospective analysis of energy security: A practical life-cycle approach focused on renewable power generation and oriented towards policy-makers, Applied Energy 190 (2017) 891–901.
[130] C. Zou, Q. Zhao, G. Zhang, B. Xiong, Energy revolution: From a fossil energy era to a new energy era, Natural Gas Industry B 3 (1) (2016) 1–11, http://www.sciencedirect.com/science/article/pii/S2352854016300109.

[131] V. Kabakian, M. McManus, H. Harajli, Attributional life cycle assessment of mounted 1.8 kwp monocrystalline photovoltaic system with batteries and comparison with fossil energy production system, Applied Energy 154 (2015) 428–437.

[132] T. Wagner, M. Themessl, A. Schuppel, A. Gobiet, H. Stigler, S. Birk, Impacts of climate change on stream flow and hydro power generation in the Alpine region, Environmental Earth Sciences 76 (1) (2016) 4–11.

[133] G. Lazzaro, G. Botter, Run-of-river power plants in Alpine regions: Whither optimal capacity?, Water Resources Research 51 (7) (2015) 5658–5676.

[134] A. Botelho, P. Ferreira, F. Lima, L.M.C. Pinto, S. Sousa, Assessment of the environmental impacts associated with hydropower, Renewable & Sustainable Energy Reviews 70 (2017) 896–904, http://www.sciencedirect.com/science/article/pii/S1364032116310462.

[135] H. Hondo, Life cycle GHG emission analysis of power generation systems: Japanese case, Energy 30 (11) (2005) 2042–2056.

[136] L. Chai, X. Liao, L. Yang, X. Yan, Assessing life cycle water use and pollution of coal-fired power generation in China using input-output analysis, Applied Energy 231 (2018) 951–958.

[137] S.H. Farjana, N. Huda, M.A.P. Mahmud, Life cycle analysis of copper-gold-lead-silver-zinc beneficiation process, Science of the Total Environment 659 (2019) 41–52.

[138] M. Marsmann, A. Schiburr, Databases, software and LCA applications, in: Environmental Life Cycle Assessment and its Applications, Journal of Cleaner Production 1 (3) (1993) 206–207, http://www.sciencedirect.com/science/article/pii/0959652693900213.

[139] V.B. Miller, A.E. Landis, L.A. Schaefer, A benchmark for life cycle air emissions and life cycle impact assessment of hydrokinetic energy extraction using life cycle assessment, Renewable Energy 36 (3) (2011) 1040–1046.

[140] C. Gabbud, S.N. Lane, Ecosystem impacts of Alpine water intakes for hydropower: the challenge of sediment management, Wiley Interdisciplinary Reviews: Water 3 (1) (2015) 41–61.

[141] X. Zhang, H.-Y. Li, Z.D. Deng, C. Ringler, Y. Gao, M.I. Hejazi, L.R. Leung, Impacts of climate change, policy and water-energy-food nexus on hydropower development, Renewable Energy 116 (2018) 827–834.

[142] Q. Zhang, B. Karney, H.L. MacLean, J. Feng, Life-cycle inventory of energy use and greenhouse gas emissions for two hydropower projects in China, Journal of Infrastructure Systems 13 (4) (2007) 271–279.

[143] E.G. Hertwich, Addressing biogenic greenhouse gas emissions from hydropower in LCA, Environmental Science & Technology 47 (17) (2013) 9604–9611.

[144] M.A.P. Mahmud, N. Huda, S.H. Farjana, C. Lang, Environmental profile evaluations of piezoelectric polymers using life cycle assessment, IOP Conference Series: Earth and Environmental Science 154 (1) (2018) 012017.

[145] S.H. Farjana, N. Huda, M.A.P. Mahmud, Environmental impact assessment of European non-ferro mining industries through life-cycle assessment, IOP Conference Series: Earth and Environmental Science 154 (1) (2018) 012019.

[146] H.A.U. de Haes, R. Heijungs, Life-cycle assessment for energy analysis and management, Applied Energy 84 (7) (2007) 817–827.

[147] S.H. Farjana, N. Huda, M.P. Mahmud, Impacts of aluminum production: A cradle to gate investigation using life-cycle assessment, Science of the Total Environment 663 (2019) 958–970.

[148] S. Žideonienė, J. Kruopienė, Life cycle assessment in environmental impact assessments of industrial projects: towards the improvement, in: Bridges for a more sustainable future: Joining Environmental Management for Sustainable Universities (EMSU) and the European Roundtable for Sustainable Consumption and Production

(ERSCP) conferences, Journal of Cleaner Production 106 (2015) 533–540, http://www.sciencedirect.com/science/article/pii/S0959652614008099.
[149] R. Dones, C. Bauer, R. Bolliger, T. Heck, A. Roder, M. Emenegger, R. Frischknecht, N. Jungbluth, M. Tuchschmid, Life cycle inventories of energy systems: results for current systems in Switzerland and other UCTE countries, Ecoinvent report no. 5, Paul Scherrer Institut, Villigen & Swiss Centre for Life Cycle Inventories, Dübendorf, Switzerland, 2007, pp. 97–106.
[150] S.H. Farjana, N. Huda, M.A.P. Mahmud, C. Lang, Comparative life-cycle assessment of uranium extraction processes, Journal of Cleaner Production 202 (2018) 666–683.
[151] M.A.J. Huijbregts, Z.J.N. Steinmann, P.M.F. Elshout, G. Stam, F. Verones, M. Vieira, M. Zijp, A. Hollander, R. van Zelm, ReCiPe2016: a harmonised life cycle impact assessment method at midpoint and endpoint level, The International Journal of Life Cycle Assessment 22 (2) (2017) 138–147.
[152] J.C. Minx, M. Callaghan, W.F. Lamb, J. Garard, O. Edenhofer, Learning about climate change solutions in the IPCC and beyond, Environmental Science & Policy 77 (2017) 252–259, http://www.sciencedirect.com/science/article/pii/S1462901117305464.
[153] A. Audenaert, S.H.D. Cleyn, M. Buyle, LCA of low-energy flats using the Eco-indicator 99 method: Impact of insulation materials, Energy and Buildings 47 (2012) 68–73.
[154] B. Truffer, J. Markard, C. Bratrich, B. Wehrli, Green electricity from Alpine hydropower plants, Mountain Research and Development 21 (1) (2001) 19–24.
[155] G.K. Heilig, The greenhouse gas methane (CH4): Sources and sinks, the impact of population growth, possible interventions, Population and Environment 16 (2) (1994) 109–137.
[156] N. Armaroli, V. Balzani, Towards an electricity-powered world, Energy & Environmental Science 4 (2011) 3193–3222, https://doi.org/10.1039/C1EE01249E.
[157] L.D. Benedetto, J. Klemeš, The environmental performance strategy map: an integrated LCA approach to support the strategic decision-making process, in: Early-Stage Energy Technologies for Sustainable Future: Assessment, Development, Application, Journal of Cleaner Production 17 (10) (2009) 900–906, http://www.sciencedirect.com/science/article/pii/S0959652609000316.
[158] E.G. Hertwich, T. Gibon, E.A. Bouman, A. Arvesen, S. Suh, G.A. Heath, J.D. Bergesen, A. Ramirez, M.I. Vega, L. Shi, Integrated life-cycle assessment of electricity-supply scenarios confirms global environmental benefit of low-carbon technologies, Proceedings of the National Academy of Sciences 112 (20) (2015) 6277–6282, http://www.pnas.org/content/112/20/6277.
[159] H.L. Raadal, L. Gagnon, I.S. Modahl, O.J. Hanssen, Life cycle greenhouse gas (GHG) emissions from the generation of wind and hydro power, Renewable & Sustainable Energy Reviews 15 (7) (2011) 3417–3422, http://www.sciencedirect.com/science/article/pii/S1364032111001924.
[160] R. Brizmohun, T. Ramjeawon, A. Azapagic, Life cycle assessment of electricity generation in Mauritius, in: Bridges for a more sustainable future: Joining Environmental Management for Sustainable Universities (EMSU) and the European Roundtable for Sustainable Consumption and Production (ERSCP) conferences, Journal of Cleaner Production 106 (2015) 565–575, http://www.sciencedirect.com/science/article/pii/S095965261401213X.
[161] J. Ling-Chin, O. Heidrich, A. Roskilly, Life cycle assessment (LCA) – from analysing methodology development to introducing an LCA framework for marine photovoltaic (PV) systems, Renewable & Sustainable Energy Reviews 59 (2016) 352–378, http://www.sciencedirect.com/science/article/pii/S1364032115014410.
[162] A. Tukker, Life cycle assessment as a tool in environmental impact assessment, Environmental Impact Assessment Review 20 (4) (2000) 435–456, http://www.sciencedirect.com/science/article/pii/S0195925599000451.

[163] F. Asdrubali, G. Baldinelli, F. D'Alessandro, F. Scrucca, Life cycle assessment of electricity production from renewable energies: Review and results harmonization, Renewable & Sustainable Energy Reviews 42 (2015) 1113–1122, http://www.sciencedirect.com/science/article/pii/S1364032114009071.

[164] M. Goralczyk, Life-cycle assessment in the renewable energy sector, Applied Energy 75 (3) (2003) 205–211.

[165] M.S. Uddin, S. Kumar, Energy, emissions and environmental impact analysis of wind turbine using life cycle assessment technique, Journal of Cleaner Production 69 (2014) 153–164, http://www.sciencedirect.com/science/article/pii/S0959652614000973.

[166] R. Garcia, P. Marques, F. Freire, Life-cycle assessment of electricity in Portugal, Applied Energy 134 (2014) 563–572, http://www.sciencedirect.com/science/article/pii/S0306261914008782.

[167] K.B. Oebels, S. Pacca, Life cycle assessment of an onshore wind farm located at the northeastern coast of Brazil, Renewable Energy 53 (2013) 60–70, http://www.sciencedirect.com/science/article/pii/S0960148112006714.

[168] Consulting project of the Chinese Academy of Engineering research on the greenhouse gas emissions of various power generation in China, Atomic Energy Press, 2015.

[169] R.S. Jorge, E.G. Hertwich, Environmental evaluation of power transmission in Norway, Applied Energy 101 (2013) 513–520.

[170] G. Finnveden, M.Z. Hauschild, T. Ekvall, J. Guinée, R. Heijungs, S. Hellweg, A. Koehler, D. Pennington, S. Suh, Recent developments in life cycle assessment, Journal of Environmental Management 91 (1) (2009) 1–21, http://www.sciencedirect.com/science/article/pii/S0301479709002345.

[171] Comparison of life cycle greenhouse gas emissions of various power sources, World Nuclear Association Report, 2011.

[172] N. Jungbluth, M. Stucki, R. Frischknecht, S. Buesser, Photovoltaics, in: R. Dones, et al. (Eds.), Sachbilanzen von Energiesystemen: Grundlagen für den ökologischen Vergleich von Energiesystemen und den Einbezug von Energiesystemen in Ökobilanzen für die Schweiz, Esu-services Ltd, Uster, CH, 2010, Ecoinvent report no. 6-xii.

[173] R. Frischknecht, M. Tuchschmid, M. Faist Emmeneger, C. Bauer, R. Dones, Strommix und Stromnetz, in: R. Dones (Ed.), Sachbilanzen von Energiesystemen: Grundlagen für den ökologischen Vergleich von Energiesystemen und den Einbezug von Energiesystemen in Ökobilanzen für die Schweiz, Paul Scherrer Institut, Villigen & Swiss Centre for Life Cycle Inventories, Dübendorf, Switzerland, 2007, Ecoinvent report no. 6.

[174] L. Wright, B. Boundy, B. Perlack, S. Davis, B. Saulsbury, Biomass Energy Data Book, ed. 1, Oak Ridge Natl. Lab., Oak Ridge, TN, 2006, ORNL/TM-2006/571.

[175] J.C. Bare, Traci, Journal of Industrial Ecology 6 (3) (2002) 49–78.

[176] K. Czaplicka-Kolarz, J. Wachowicz, M. Bojarska-Kraus, A life cycle method for assessment of a colliery's Eco-indicator, The International Journal of Life Cycle Assessment 9 (4) (2004) 247–253.

[177] S. Suh, M. Lenzen, G.J. Treloar, H. Hondo, A. Horvath, G. Huppes, O. Jolliet, U. Klann, W. Krewitt, Y. Moriguchi, J. Munksgaard, G. Norris, System boundary selection in life-cycle inventories using hybrid approaches, Environmental Science & Technology 38 (3) (2004) 657–664, https://doi.org/10.1021/es0263745, pMID: 14968848.

[178] F. Pomponi, M. Lenzen, Hybrid life cycle assessment (LCA) will likely yield more accurate results than process-based LCA, Journal of Cleaner Production 176 (2018) 210–215, http://www.sciencedirect.com/science/article/pii/S0959652617330780.

[179] T.O. Wiedmann, S. Suh, K. Feng, M. Lenzen, A. Acquaye, K. Scott, J.R. Barrett, Application of hybrid life cycle approaches to emerging energy technologies – the

case of wind power in the UK, Environmental Science & Technology 45 (13) (2011) 5900–5907, https://doi.org/10.1021/es2007287, pMID: 21649442.
[180] A.H. Strømman, Dealing with double-counting in tiered hybrid life-cycle inventories: a few comments – response, Journal of Cleaner Production 17 (17) (2009) 1607–1609, http://www.sciencedirect.com/science/article/pii/S0959652609002108.
[181] M. Lenzen, R. Crawford, The path exchange method for hybrid LCA, Environmental Science & Technology 43 (21) (2009) 8251–8256, https://doi.org/10.1021/es902090z, pMID: 19924952.
[182] M. Lenzen, J. Munksgaard, Energy and CO_2 life-cycle analyses of wind turbines—review and applications, Renewable Energy 26 (3) (2002) 339–362, http://www.sciencedirect.com/science/article/pii/S0960148101001458.
[183] A. Malik, M. Lenzen, A. Geschke, Triple bottom line study of a lignocellulosic biofuel industry, GCB Bioenergy 8 (1) (2016) 96–110, https://onlinelibrary.wiley.com/doi/abs/10.1111/gcbb.12240.
[184] M. Lenzen, Greenhouse gas analysis of solar-thermal electricity generation, Solar Energy 65 (6) (1999) 353–368, http://www.sciencedirect.com/science/article/pii/S0038092X99000055.
[185] A. Malik, M. Lenzen, P.J. Ralph, B. Tamburic, Hybrid life-cycle assessment of algal biofuel production, in: Advances in biofuels and chemicals from algae, Bioresource Technology 184 (2015) 436–443, http://www.sciencedirect.com/science/article/pii/S0960852414015648.
[186] H. Imbeault-Tétreault, O. Jolliet, L. Deschênes, R.K. Rosenbaum, Analytical propagation of uncertainty in life cycle assessment using matrix formulation, Journal of Industrial Ecology 17 (4) (2013) 485–492, https://onlinelibrary.wiley.com/doi/abs/10.1111/jiec.12001.
[187] R. Crawford, G. Treloar, R. Fuller, M. Bazilian, Life-cycle energy analysis of building integrated photovoltaic systems (BiPVs) with heat recovery unit, Renewable & Sustainable Energy Reviews 10 (6) (2006) 559–575, http://www.sciencedirect.com/science/article/pii/S1364032105000055.
[188] N.M. Kumar, S.S. Chopra, M. Malvoni, R.M. Elavarasan, N. Das, Solar cell technology selection for a PV leaf based on energy and sustainability indicators—a case of a multilayered solar photovoltaic tree, Energies 13 (23) (2020).
[189] T.N. Ligthart, R.H. Jongbloed, J.E. Tamis, A method for improving centre for environmental studies (CML) characterisation factors for metal (eco)toxicity — the case of zinc gutters and downpipes, The International Journal of Life Cycle Assessment 15 (8) (Sep 2010) 745–756, https://doi.org/10.1007/s11367-010-0208-z.
[190] M.P. Mahmud, N. Huda, S.H. Farjana, C. Lang, Life-cycle impact assessment of renewable electricity generation systems in the United States, Renewable Energy 151 (2020) 1028–1045.
[191] A. Anvari-Moghaddam, J.M. Guerrero, J.C. Vasquez, H. Monsef, A. Rahimi-Kian, Efficient energy management for a grid-tied residential microgrid, IET Generation, Transmission & Distribution 11 (2017) 2752–2761.
[192] M. You, J. Jiang, A.M. Tonello, T. Doukoglou, H. Sun, On statistical power grid observability under communication constraints, IET Smart Grid 1 (2018) 40–47.
[193] D. Fioriti, R. Giglioli, D. Poli, G. Lutzemberger, A. Vanni, P. Salza, Optimal sizing of a mini-grid in developing countries, taking into account the operation of an electrochemical storage and a fuel tank, in: 2017 6th International Conference on Clean Electrical Power (ICCEP), vol. 3, 2017, pp. 320–326.
[194] M. Lee, G.C. Shaw, V. Modi, Battery storage: Comparing shared to individually owned storage given rural demand profiles of a cluster of customers, in: IEEE Global Humanitarian Technology Conference (GHTC 2014), vol. 2, 2014, pp. 200–206.

[195] R.K. Sharma, S. Mishra, Dynamic power management and control of a PV PEM fuel-cell-based standalone ac/dc microgrid using hybrid energy storage, IEEE Transactions on Industry Applications 54 (1) (2018) 526–538.

[196] Y. Karimi, H. Oraee, M.S. Golsorkhi, J.M. Guerrero, Decentralized method for load sharing and power management in a PV/battery hybrid source islanded microgrid, IEEE Transactions on Power Electronics 32 (5) (2017) 3525–3535.

[197] C. Liu, X. Wang, X. Wu, J. Guo, Economic scheduling model of microgrid considering the lifetime of batteries, IET Generation, Transmission & Distribution 11 (2017) 759–767.

[198] M. Rezkallah, A. Hamadi, A. Chandra, B. Singh, Design and implementation of active power control with improved P&O method for wind-PV-battery-based standalone generation system, IEEE Transactions on Industrial Electronics 65 (7) (2018) 5590–5600.

[199] S. Shrivastava, B. Subudhi, S. Das, Distributed voltage and frequency synchronisation control scheme for islanded inverter-based microgrid, IET Smart Grid 1 (2018) 48–56.

[200] N. Liu, M. Cheng, X. Yu, J. Zhong, J. Lei, Energy-sharing provider for PV prosumer clusters: A hybrid approach using stochastic programming and Stackelberg game, IEEE Transactions on Industrial Electronics 65 (8) (2018) 6740–6750.

[201] S.H. Farjana, N. Huda, M.A.P. Mahmud, S. Rahman, Solar industrial process heating systems in operation-current SHIP plants and future prospects in Australia, Renewable & Sustainable Energy Reviews 91 (2018) 409–419.

[202] P. Prabhakaran, Y. Goyal, V. Agarwal, Novel nonlinear droop control techniques to overcome the load sharing and voltage regulation issues in dc microgrid, IEEE Transactions on Power Electronics 33 (5) (2018) 4477–4487.

[203] M.F. Shaaban, A.H. Osman, M.S. Hassan, Day-ahead optimal scheduling for demand side management in smart grids, in: 2016 European Modelling Symposium (EMS), vol. 3, 2016, pp. 124–129.

[204] D. Ioli, A. Falsone, M. Prandini, An iterative scheme to hierarchically structured optimal energy management of a microgrid, in: 2015 54th IEEE Conference on Decision and Control (CDC), vol. 2, 2015, pp. 5227–5232.

[205] A. Meliopoulos, Challenges in simulation and design of microgrids, in: 2002 IEEE Power Engineering Society Winter Meeting, vol. 1, 2002, pp. 309–314.

[206] M.A. Kabir, A.S.M.M. Hasan, T.H. Sakib, S.J. Hamim, Challenges of photovoltaic based hybrid minigrid for off-grid rural electrification in Bangladesh, in: 2017 4th International Conference on Advances in Electrical Engineering (ICAEE), vol. 2, 2017, pp. 686–690.

[207] Q. Sun, J. Zhou, J.M. Guerrero, H. Zhang, Hybrid three-phase single-phase microgrid architecture with power management capabilities, IEEE Transactions on Power Electronics 30 (10) (2015) 5964–5977.

[208] A. Bidram, A. Davoudi, F.L. Lewis, J.M. Guerrero, Distributed cooperative secondary control of microgrids using feedback linearization, IEEE Transactions on Power Systems 28 (3) (2013) 3462–3470.

[209] J. Ni, L. Liu, C. Liu, X. Hu, S. Li, Secondary voltage control for microgrids based on fixed-time distributed cooperative control of multi-agent systems, in: 2017 American Control Conference (ACC), vol. 3, 2017, pp. 761–766.

[210] J.G. de Matos, L.A.d.S. Ribeiro, E.d.C. Gomes, Power control in isolated Microgrids with renewable distributed energy sources and battey banks, in: 2013 International Conference on Renewable Energy Research and Applications (ICRERA), vol. 2, 2013, pp. 258–263.

[211] D.R. Quete, C.A. Canizares, An affine arithmetic-based energy management system for isolated microgrids, IEEE Transactions on Smart Grid 99 (2018) 1–12.

[212] A. Urtasun, E.L. Barrios, P. Sanchis, L. Marroyo, Frequency-based energy-management strategy for stand-alone systems with distributed battery storage, IEEE Transactions on Power Electronics 30 (9) (2015) 4794–4808.

[213] C. Qin, P. Ju, F. Wu, Y. Jin, Q. Chen, L. Sun, A coordinated control method to smooth short-term power fluctuations of hybrid offshore renewable energy conversion system (HORECS), in: 2015 IEEE Eindhoven PowerTech, vol. 2, 2015, pp. 1–5.

[214] H. Mahmood, D. Michaelson, J. Jiang, Decentralized power management of a PV/battery hybrid unit in a droop-controlled islanded microgrid, IEEE Transactions on Power Electronics 30 (12) (2015) 7215–7229.

[215] M.C. Chandorkar, D.M. Divan, R. Adapa, Control of parallel connected inverters in standalone ac supply systems, IEEE Transactions on Industry Applications 29 (1) (1993) 136–143.

[216] M.S. Rahman, M.J. Hossain, J. Lu, Utilization of parked EV-ESS for power management in a grid-tied hybrid ac/dc microgrid, in: 2015 Australasian Universities Power Engineering Conference (AUPEC), vol. 2, 2015, pp. 1–6.

[217] H. Zhou, T. Bhattacharya, D. Tran, T.S.T. Siew, A.M. Khambadkone, Composite energy storage system involving battery and ultracapacitor with dynamic energy management in microgrid applications, IEEE Transactions on Power Electronics 26 (3) (2011) 923–930.

[218] E. Fernandez, M. Hossain, M. Nizami, Game-theoretic approach to demand-side energy management for a smart neighbourhood in Sydney incorporating renewable resources, Applied Energy 232 (2018) 245–257.

[219] M.S.H. Nizami, M.J. Hossain, K. Mahmud, J. Ravishankar, Energy cost optimization and DER scheduling for unified energy management system of residential neighborhood, in: 2018 IEEE International Conference on Environment and Electrical Engineering and 2018 IEEE Industrial and Commercial Power Systems Europe (EEEIC / I CPS Europe), 2018, pp. 1–6.

[220] M.N. Akter, M.A. Mahmud, A.M.T. Oo, A hierarchical transactive energy management system for energy sharing in residential microgrids, Energies 10 (12) (2017) 2098.

[221] V. Astapov, S. Trashchenkov, Design and reliability evaluation of standalone microgrid, in: 2017 18th International Scientific Conference on Electric Power Engineering (EPE), vol. 2, 2017, pp. 1–6.

[222] J.M. Guerrero, J.C. Vasquez, J. Matas, L.G. de Vicuna, M. Castilla, Hierarchical control of droop-controlled AC and DC microgrids: A general approach toward standardization, IEEE Transactions on Industrial Electronics 58 (2011) 158–172.

[223] M.S. Rahman, J. Hossain, J. Lu, H.R. Pota, A need-based distributed coordination strategy for EV storages in a commercial hybrid AC/DC microgrid with an improved interlinking converter control topology, IEEE Transactions on Energy Conversion 99 (2017) 1–12.

[224] M.S. Rahman, M. Hossain, J. Lu, Coordinated control of three-phase AC and DC type EV–ESSs for efficient hybrid microgrid operations, Energy Conversion and Management 122 (2016) 488–503.

[225] J. Clairand, M. Arriaga, C.A. Canizares, C. Alvarez, Power generation planning of Galapagos' microgrid considering electric vehicles and induction stoves, IEEE Transactions on Sustainable Energy (2018) 1–8.

[226] M.A. Hossain, H.R. Pota, S. Squartini, A.F. Abdou, Modified PSO algorithm for real-time energy management in grid-connected microgrids, Renewable Energy 136 (2019) 746–757.

[227] P. Kou, D. Liang, L. Gao, Distributed coordination of multiple PMSGs in an islanded DC microgrid for load sharing, IEEE Transactions on Energy Conversion 32 (2) (2017) 471–485.

[228] Adoption of the Paris Agreement, Documents FCCC/Cp/2015/L.9/Rev.1, UNFCCC, 2015.
[229] T. Givler, P. Lilienthal, Using HOMER software, NREL's micro power optimization model, to explore the role of gen-sets in small solar power systems case study: Sri Lanka, Technical Report NREL/TP-710-36774. Available from: https://www.osti.gov/biblio/15016073, https://doi.org/10.2172/15016073, 2005.
[230] K. Lau, M. Yousof, S. Arshad, M. Anwari, A. Yatim, Performance analysis of hybrid photovoltaic/diesel energy system under Malaysian conditions, Energy 35 (8) (2010) 3245–3255.
[231] HOMER analysis, https://analysis.nrel.gov/homer/. (Accessed 3 March 2019).
[232] T. Lambert, P. Gilman, P. Lilienthal, Micropower system modeling with HOMER, Nat. Renew. Energy Lab., 2006, pp. 379–385.
[233] S. Mizani, A. Yazdani, Optimal design and operation of a grid-connected microgrid, in: 2009 IEEE Electrical Power Energy Conference (EPEC), 2009, pp. 1–6.
[234] D.-R. Thiam, Renewable decentralized in developing countries: Appraisal from microgrids project in Senegal, Renewable Energy 35 (8) (2010) 1615–1623.
[235] B. Kroposki, G. Martin, Hybrid renewable energy and microgrid research work at NREL, in: IEEE PES General Meeting, 2010, pp. 1–4.
[236] P. Diaz, C. Arias, R. Pena, D. Sandoval, Far from the grid: A rural electrification field study, Renewable Energy 35 (12) (2010) 2829–2834.
[237] S. Chakraborty, M.G. Simoes, PV-microgrid operational cost minimization by neural forecasting and heuristic optimization, in: 2008 IEEE Industry Applications Society Annual Meeting, 2008, pp. 1–8.
[238] O. Hafez, K. Bhattacharya, Optimal planning and design of a renewable energy based supply system for microgrids, Renewable Energy 45 (2012) 7–15.
[239] C. Nayar, M. Tang, W. Suponthana, Wind/PV/diesel micro grid system implemented in remote islands in the Republic of Maldives, in: 2008 IEEE International Conference on Sustainable Energy Technologies, 2008, pp. 1076–1080.
[240] M. Anyi, B. Kirke, S. Ali, Remote community electrification in Sarawak, Malaysia, Renewable Energy 35 (7) (2010) 1609–1613.
[241] Y. Himri, A.B. Stambouli, B. Draoui, S. Himri, Techno-economical study of hybrid power system for a remote village in Algeria, Energy 33 (7) (2008) 1128–1136.
[242] E. Nfah, J. Ngundam, M. Vandenbergh, J. Schmid, Simulation of off-grid generation options for remote villages in Cameroon, Renewable Energy 33 (5) (2008) 1064–1072.
[243] G. Bekele, B. Palm, Feasibility study for a standalone solar–wind-based hybrid energy system for application in Ethiopia, Applied Energy 87 (2) (2010) 487–495.
[244] M.A.P. Mahmud, M. Hossain, M.S.H. Nizami, M.S. Rahman, S.H. Farjana, N. Huda, C. Lang, Advanced power routing framework for optimal economic operation and control of solar photovoltaic-based islanded microgrid, IET Smart Grid 2 (2019) 242–249.
[245] D.A. Notter, M. Gauch, R. Widmer, P. Wager, A. Stamp, R. Zah, H.-J. Althaus, Contribution of Li-ion batteries to the environmental impact of electric vehicles, Environmental Science & Technology 44 (2010) 6550–6556.
[246] H. Hao, Z. Mu, S. Jiang, Z. Liu, F. Zhao, GHG emissions from the production of lithium-ion batteries for electric vehicles in China, Sustainability 9 (4) (2017) 504.
[247] L.A.-W. Ellingsen, G. Majeau-Bettez, B. Singh, A.K. Srivastava, L.O. Valøen, A.H. Strømman, Life cycle assessment of a lithium-ion battery vehicle pack, Journal of Industrial Ecology 18 (1) (2014) 113–124.
[248] A. Prasai, A. Paquette, Y. Du, R. Harley, D. Divan, Minimizing emissions in microgrids while meeting reliability and power quality objectives, in: Power Electronics Conference (IPEC), 2010 International, IEEE, 2010, pp. 726–733.

[249] J. Bare, P. Hofstetter, D.W. Pennington, H.A. Udo de Haes, Midpoints versus endpoints: The sacrifices and benefits, The International Journal of Life Cycle Assessment 5 (2012) 319–326.

[250] M. Goedkoop, M. Oele, M. Vieira, J. Leijting, T. Ponsioen, E. Meijer, SimaPro Tutorial, vol. 89, https://network.simapro.com/, 2014.

[251] M.R. Sandgani, S. Sirouspour, Priority-based microgrid energy management in a network environment, IEEE Transactions on Sustainable Energy 9 (2) (2018) 980–990.

[252] Z. Shi, H. Liang, S. Huang, V. Dinavahi, Distributionally robust chance-constrained energy management for islanded microgrids, IEEE Transactions on Smart Grid 10 (2) (2019) 2234–2244.

[253] V. Sarfi, H. Livani, An economic-reliability security-constrained optimal dispatch for microgrids, IEEE Transactions on Power Systems 33 (2018) 6777–6786.

[254] A.A. Bashir, M. Pourakbari-Kasmaei, J. Contreras, M. Lehtonen, A novel energy scheduling framework for reliable and economic operation of islanded and grid-connected microgrids, Electric Power Systems Research 171 (2019) 85–96.

[255] A.M. Aly, A.M. Kassem, K. Sayed, I. Aboelhassan, Design of microgrid with flywheel energy storage system using HOMER software for case study, in: 2019 International Conference on Innovative Trends in Computer Engineering (ITCE), 2019, pp. 485–491.

[256] NASA Surface Meteorology and Solar Energy, http://eosweb.larc.nasa.gov/sse/. (Accessed 3 March 2019).

[257] R. Hischier, M. Classen, M. Lehmann, W. Scharnhorst, Life cycle inventories of electric and electronic equipment: production, use and disposal, Tech. Rep. 18, Swiss Federal Laboratories Mater. Sci. Technol., Dübendorf, Switzerland, 2007.

Index

A

Abiotic
  depletion, 27, 41, 96, 97, 135, 147, 148
  resource depletion impacts, 24
Acidification
  aquatic, 15, 56, 75, 91
  indicators, 30
  terrestrial, 75, 90, 182, 191, 200
Alpine
  areas, 3, 73, 77, 78, 84, 88, 91–94, 97, 206
  countries, 78
  hydropower plants, 74, 75, 78, 80, 81, 88, 93, 94, 98, 102
  plants, 24, 88, 93, 94, 102
  regions, 6, 73, 74, 78, 84, 89, 91, 94, 98
  zones, 88, 93, 95
Alternating current (AC), 2
Amount
  energy, 112, 139
  excess energy, 180, 185
  fossil fuel, 87
Aquatic
  acidification, 15, 56, 75, 91
  ecosystems, 192
  ecotoxicity, 15, 41, 56, 74, 75, 91, 96, 135, 147, 148
  eutrophication, 15, 42, 56, 75, 91
Assessing
  climate change, 151
  environmental profiles, 6
  GHG emissions, 86, 146
Assessment
  for GHG emission, 4
  indicators, 7, 131, 205
  methods, 14, 54

B

Battery
  energy storage, 72, 162
  lifetimes, 199
  storages, 49, 58, 181
  strings, 187

Battery energy storage system (BESS), 2, 161–163
Biogenic methane, 85
Biomass
  application for electricity production, 27
  availability, 140
  combustion processes, 40
  decay, 96
  energy, 120
  for electricity production, 27
  gasification, 33
  plants, 11, 17, 35, 41–43, 103, 105, 118, 121, 131, 135, 153, 156, 205
  power, 94
  power plants, 27, 41, 42, 104, 105, 111, 118–121, 127, 133, 136–140, 144, 147–150, 153, 154, 160, 207, 208
  power plants environmental impacts, 9
  technologies, 29
Bituminous coal plants, 133

C

Carbon
  dioxide emissions, 31, 48, 65, 74, 102, 182
  dioxide gas, 93
  emissions, 9
  emissions renewable energy, 10
Carbon monoxide emissions, 86
Carcinogenics, 103, 116, 118, 119, 124, 127, 133
Categories
  carbon dioxide, 93
  environmental, 182
  hydropower plants, 77
  plants, 4, 5, 77, 206
Central storage system (CSS), 161, 162, 164, 207
Clean
  energy, 47, 73
  hydropower generation, 102
  renewable energy, 6
  renewable power production, 43

231

Climate change, 3, 15, 24, 48, 56, 59, 61, 64, 68, 73, 75, 77, 86, 91, 129, 192, 206, 207
　assessing, 151
　factors, 58
Coal power plants, 155
Community storage
　lifetimes, 199
　size, 185
　units, 184, 202
　units environmental profiles, 198
Comparative life cycle inputs, 93
Comprehensive
　LCA analysis, 39
　LCI, 49, 55, 102, 104, 105, 137, 138, 206
Consumed energy, 118, 147
Control operation
　method, 6
　ordered, 168
　stable, 161, 163, 178, 205, 207
　strategy, 164
Controller operation, 175, 176
Copper indium selenide (CIS), 47, 68, 183, 189, 193, 207
Cost of energy (COE), 180, 181, 185
Cumulative energy demand (CED), 4, 6, 15, 47, 73, 103, 135, 206

D

Damage indicators, 40
Dangerous
　elements, 202, 209
　emissions, 74
　GHGs, 73
Direct current (DC), 2
　bus, 163, 165, 169, 176, 184
　bus voltages, 176
　converters, 184
Disposal amounts, 115, 140

E

Ecoinvent database, 55, 85, 105, 115, 123, 137, 144, 145, 147, 182, 189, 207
Ecological impacts, 49, 109, 139
Economic analysis (EA), 5
Economic inputs, 2, 4

Ecosystems
　aquatic, 192
　freshwater, 88, 191
　quality, 56, 75, 86, 91, 92, 98, 120, 124, 133, 148, 160, 207
　terrestrial, 88, 191, 192
Ecotoxicity
　aquatic, 15, 41, 56, 74, 75, 91, 96, 135, 147, 148
　categories, 127
　freshwater, 63, 72, 96, 97, 148, 194, 200
　terrestrial, 15, 42, 56, 75, 91, 136, 147, 148, 154, 194
Effect indicators, 22, 40, 138
Effect indicators environmental, 39, 193
Electrical energy amount, 82
Electricity
　cogeneration, 31
　consumption, 18, 156
　demand, 1, 7, 184
　generation, 27, 31, 109, 156, 184
　　from biomass, 98
　　from wind, 98
　　sensitivity analysis, 29
　　sustainability assessment, 35
　　systems, 10, 135
　　technologies, 10, 135
　grid, 33, 35, 37
　market, 27
　outcomes, 132
　production, 6, 7, 13, 27, 31, 78, 84, 88, 115, 120, 140, 148, 180, 205
　renewable, 1, 23
　sources, 109
Elements
　dangerous, 202, 209
　environmental impacts, 49
　hydropower plants, 84
　manufacturing, 49, 71, 209
　optimal sizing, 4, 183
　plants, 18, 84, 87, 104, 115, 137
　renewable energy, 205
　renewable power plants, 41
Embodied energy, 57, 149
Emissions
　dangerous, 74
　from renewable energy plants, 11

from RETs, 104, 136
hazardous, 71–74, 209
rates, 93, 104, 122, 150
End-of-life recycling quantities, 115, 140
End-point environmental impacts, 91
Energy
  amount, 112, 139
  balance constraints, 168
  biomass, 120
  consumption, 55, 87, 94, 108, 109, 129, 138, 145, 182
  consumption amount, 66
  consumption rates, 78
  cost, 163
  crisis, 48
  demands, 163
  flow, 118, 147
  generation renewable, 24
  generation scheduling, 162
  intake, 189
  losses, 190
  management, 163, 181
  meter, 59, 189, 194–196, 198
  nuclear, 109, 121, 133, 149, 150, 208
  output, 189
  production, 47, 54, 81, 109, 131
  rates, 81, 165
  relative amounts, 139
  renewable, 1, 5, 9, 11, 31, 35, 47, 87, 95, 103, 111, 135, 139, 145, 165
  sharing, 161, 162, 165, 166
  solar, 47, 52, 135
  sources, 109, 180
  storage, 161, 172
  storage device, 69
  storage unit, 179
  transmission distances, 180
  usage, 87, 115, 145
  wind, 41, 104, 109, 136
Energy payback time (EPBT), 22, 23, 27
Environmental
  categories, 182
  effect, 3, 6, 23, 26, 27, 33, 35, 48, 58, 61, 77, 80, 85, 103, 108, 109, 135, 139, 147, 182, 205, 206
  effect indicators, 39, 193
  hazards, 3, 10, 11, 33, 54, 61, 77, 87, 118, 147, 160, 206, 209

impact, 1–4, 7, 8, 11, 15, 23, 24, 47–49, 56, 80, 85, 91, 96, 97, 103–105, 135–139, 146, 147, 180, 182, 183, 188, 194, 206–208
  assessment, 4, 5, 118, 183, 185, 188, 194, 208
  calculations, 152, 153
  comparison, 118, 147
  estimation methods, 7
  evaluation, 8, 188
  indicators, 58, 188
  outcome, 147
  performance, 3, 6, 23, 24, 27, 46, 47, 50, 66, 72, 201
  profiles, 3, 71, 78, 89, 198
  profiles assessing, 6
  sustainability assessment, 24
Europe
  alpine regions, 8
  hydropower plants, 3, 75, 140
  nonalpine areas, 3, 5, 73, 74, 77, 80, 84, 206
  nonalpine regions, 8, 74, 88
Eutrophication
  aquatic, 15, 42, 56, 75, 91
  categories, 127
  freshwater, 59, 61, 90, 182, 191, 194, 195
  terrestrial, 41, 61, 63, 72
Excess electricity, 2, 181, 184, 193
Excess energy, 163, 167, 174, 176, 178, 181, 183, 193, 205, 207
  amount, 180, 185
  for routing, 175, 208
  routing, 174

F

Flow rates, 54
Forest biomass chains, 30
Fossil fuels
  amount, 87, 121
  combustion rates, 97
  consumption, 41, 66, 87, 117, 121, 145, 180, 195
  consumption rates, 40, 41
  depletion, 103, 116, 118, 119, 133
  depletion categories, 124
  usage, 150

Framework installation, 54
Freshwater
　ecosystems, 88, 191
　ecotoxicity, 63, 72, 96, 97, 148, 194, 200
　eutrophication, 59, 61, 90, 182, 191, 194, 195
Fuel inputs, 57, 120, 149
Fulfilling prosumers, 6, 164
Functional unit, 11, 13, 54, 56, 81, 82, 112, 139, 188

G

Greenhouse gas (GHG), 1, 8, 24, 41, 48, 65, 75, 117, 122, 123, 146, 151, 192, 198, 206, 207
　emissions, 2, 8, 10, 23, 24, 47, 49, 65, 73, 74, 86, 93, 96, 104, 105, 108, 117, 122, 137, 138, 146, 150, 151, 179, 180, 182, 196, 205, 208
　emissions assessing, 86, 146
　emissions rates, 3, 22, 23, 45, 58, 71, 96, 97
　emissions renewable power plants, 40
Grid electricity, 31

H

Harmful emissions, 37, 59
Hazardous
　emissions, 71–74, 209
　emissions from hydropower plants, 23
　gas emissions, 58, 150
　materials, 5, 18, 49, 61, 63, 148
　materials in plant, 46
　materials in plant installations, 102
　metals, 138, 206
　raw material amount, 104, 137
Heat transfer fluid (HTF) tank, 49, 52, 61, 63, 65
Higher heating values (HHV), 149
Human
　damage indicators, 86
　health, 1, 3, 10, 15, 26, 42, 56, 59, 61, 64, 73–75, 77, 86, 88, 103, 117, 124, 133, 148, 149, 160, 179, 191–193, 206, 207
　health impact categories, 92
　health indicator, 88
　toxicity, 24, 27, 35, 59, 61, 64, 68, 72, 86, 96, 97, 136, 147, 148
Hybrid operation, 161
Hybrid Optimization Model for Electric Renewables (HOMER), 180
Hydro turbines, 147
Hydroelectric power
　generation, 74
　system, 23
　technologies, 37
Hydroelectricity generation, 81
Hydrogen energy carriers, 31
Hydropower
　assets, 140
　generation enormous amount, 73, 74
　generation systems, 1
　installations, 135
　plants, 1, 3–6, 8–11, 23, 24, 35, 41, 73–75, 78, 84, 85, 87, 88, 90, 92–94, 96, 97, 104, 136–139, 147–151, 205–208
　categories, 77
　elements, 84
　environmental impacts, 3, 5, 73, 74
　environmental profiles, 88
　GHG emission, 156
　GHG net reservoir emissions, 39
　LCA analysis, 41
　sensitivity analysis, 24
　production, 78, 84, 102
　systems, 24
　technologies environmental impacts, 9

I

Impactful elements, 11
Impacts
　calculations environmental, 152, 153
　comparison environmental, 118, 147
　environmental, 1–4, 7, 8, 11, 15, 23, 24, 47–49, 56, 80, 85, 91, 96, 97, 103–105, 135–139, 146, 147, 180, 182, 183, 188, 194, 206–208
　evaluation environmental, 188
　human health, 120
　production, 5
　renewable power plants, 40, 41
Indicators assessment, 7, 131, 205

Indirect impacts, 1
Input
  economic, 2, 4, 183
  energies, 112, 140
  lifetime, 78
  raw materials, 112, 140
  resources, 112, 140
Installation
  capacity, 24
  plants, 111, 139
Intelligent energy management, 163
Interconnecting inverter, 176
Intergovernmental Panel on Climate
    Change (IPCC), 4, 15, 47, 73, 75,
    81, 86, 103, 146, 182, 190, 206
  approach, 58, 65, 131, 179, 182, 192
  for GHG emission, 72, 202, 207
  methods, 6, 40, 86, 105, 117, 118, 122,
    131, 132, 147, 151, 160, 183, 196,
    206
International Life Cycle Data System
    (ILCD), 4, 15, 47, 72, 132, 206
  approach, 61
  method, 40, 56, 58, 59, 63
International Standardization Organization
    (ISO), 8, 54, 74, 109, 139, 188
Inverter controller, 169–171
Ionizing radiation, 15, 42, 56, 59, 64, 68,
    72, 73, 75, 90, 91, 98, 153, 182,
    192, 194, 195, 200, 205, 206
Islanded
  MG, 4, 161, 162, 179, 205
  MG framework, 3, 5
  MG operation, 162
  MG systems, 181
  operation, 162

L

Land
  occupation, 15, 42, 56, 75, 91
  use, 40, 56, 59, 61, 63, 72, 73, 90, 129,
    182, 192, 194, 202, 205, 206, 208
Life cycle
  emissions, 3, 4
  environmental impacts, 1, 4, 31, 54, 63,
    72, 81, 110, 139
  impacts, 3, 5

input, 58, 189, 206
input–output rates, 61
Life cycle assessment (LCA), 1, 7, 8, 47, 73,
    103, 136, 179, 206
  analysis, 3, 13, 23, 30, 31, 33, 35, 37, 39,
    45, 49, 54, 56, 58, 71, 74, 102, 115,
    117, 140, 179, 182, 183, 188
  approach, 1, 8, 54, 80, 109, 139, 156,
    182
  approaches, 11
  impacts, 14
  renewable power plants, 12
Life cycle inventory (LCI), 1, 13, 47, 74,
    104, 118, 135–137, 147, 179, 205
Lifetime
  cost feasibility, 187
  environmental impacts, 27, 185
  inputs, 78
  renewable power plants, 2, 40
Local
  energy sources, 183
  power routing, 162
Lower heating value (LHV), 149

M

Manufacturing
  elements, 49, 71, 209
  phases, 121
  processes, 110, 139
  step, 109, 139
Marine
  ecotoxicity, 195
  eutrophication, 59
Metal particle emissions, 198
Methane
  biogenic emission, 88, 92, 94
  biogenic rates, 92
  potential release, 96
Microgrid controller (MGC), 165
Microgrid (MG), 2, 161–163, 165, 179,
    180, 198, 205
  bus, 163
  components, 179
  effectiveness, 164, 179
  framework, 2, 162–165, 171–174, 180
  inverter, 165
  operation, 169, 189

power quality, 165
power-routing strategy, 168
system, 162, 180
system overview, 183
Microhydropower installations, 24
Mineral extraction, 75
Mini hydropower plant, 74
Monocrystalline PV modules, 18

N

National grid, 178, 180, 181, 185, 187, 202, 209
National Renewable Energy Laboratory (NREL), 180
Net energy savings, 137
Net present cost (NPC), 2–4, 179, 207
Nitrous oxide emission, 65
Nonalpine
  areas, 73, 75, 78, 81, 91, 93, 94, 97
  hydropower plants, 94, 98, 102
  plants, 73, 93, 94, 98
  plants methane release, 102
  regions, 73, 78, 84, 88, 92, 98
  zones, 81, 88
Noncarcinogenics, 116, 118, 119, 127
Noncarcinogenics emissions, 86
Noncarcinogens, 15, 42, 56, 75, 91, 98, 124
Nonfunctional units, 145
Nonlinear programming (NLP), 2, 4, 5, 162
Nonrenewable
  electricity generation, 26, 135
  energy, 56, 75, 84, 115, 140
  energy sources, 135, 209
Nuclear
  energy, 109, 121, 133, 149, 150, 208
  plants, 75, 131
  power plants, 124

O

Occupation impacts, 40
Off-grid remote MG framework, 164
Offshore wind
  plants, 27
  power technologies, 26

Onshore wind power, 27
Operation
  islanded, 162
  optimal, 193
  phase, 97
  sustainable, 187
Operational constraints, 162, 167
Optimal
  operation, 193
  operation strategy, 187
  power routing, 166, 168, 174
  sizing, 2, 3, 179, 183, 187
Output emissions, 112, 140
Ozone
  depletion, 59, 61, 68, 73, 103, 116, 118, 124, 127, 133, 205, 206
  formation, 73, 85, 88, 182, 191, 194, 205, 206
  layer depletion, 15, 42, 48, 56, 75, 91, 98, 133, 135, 147, 148, 208

P

Photochemical ozone depletion, 59
Photovoltaic (PV), 1, 8, 47, 103, 135, 161, 179
  generation, 137, 164–169, 173–175, 178, 205, 207
  generation units, 171
  installations, 23
  modules, 18, 33, 41, 46, 179, 180, 182, 183, 195, 199
  panels, 5, 9, 18, 23, 48, 49, 52, 58, 59, 66, 98, 102, 104, 137, 147, 161, 166, 176, 184, 187, 189, 193, 195, 196
  power generation, 165, 174
  power plants, 135
Photovoltaic thermal (PVT) system, 18
Plants
  alpine, 24, 88, 93, 94, 102
  area, 140
  biomass, 11, 17, 35, 41, 42, 103, 105, 118, 121, 131, 135, 153, 156, 205
  categories, 4, 5, 77, 206
  construction, 97
  design, 74
  elements, 18, 84, 87, 104, 115, 137

Index   237

elements manufacturing, 105, 138–140, 206
elements transportation distances, 97
hydropower, 1, 3–6, 8–11, 23, 24, 35, 41, 73–75, 78, 84, 85, 87, 92–94, 96, 97, 104, 136–139, 147–151, 205–208
infrastructure, 84
installation, 111, 139
locations, 55, 88, 104
nonalpine, 73, 93, 94, 98
nuclear, 75, 131
position, 131
power, 1, 6–8, 22, 31, 41, 46, 78, 98, 103, 105, 123, 124, 131, 135, 136, 152, 155, 209
renewable, 1, 2, 6, 9, 11, 31, 40–42, 103, 129, 156
renewable energy, 1, 5, 8, 115, 122, 140, 150
renewable power, 1, 4, 10, 14, 17, 31, 40, 41, 104, 109, 116, 118, 121, 136, 138, 146, 147, 150, 180, 206, 207
solar, 18, 42, 55, 154
structure, 120, 148
types, 14
wind, 27, 208
wind power, 24, 26, 27, 135–140, 147–149, 151, 153, 207
Power
biomass, 94
generation, 31, 74, 104, 124, 136, 168
generation units, 104
plant, 1, 6–8, 22, 31, 41, 46, 78, 98, 103, 105, 123, 124, 131, 135, 136, 152, 155, 209
plant elements, 209
plant impacts, 8
production plant, 75
solar, 17
wind, 37
Power routing, 6, 161, 162, 164, 165, 178, 207
Pumped storage hydropower, 105, 109, 118, 119, 123, 131, 133
plants, 3, 4, 6, 103, 105, 108, 111, 115, 118–124, 127, 129, 206, 208

production systems, 109
Pumped storage plants, 102

R

Rates
energy, 81, 165
methane biogenic, 92
Raw material, 7, 27, 41, 55, 84, 93, 97, 104, 115, 116, 136, 140, 147
extraction, 2, 7, 13, 24, 54, 55, 75, 87, 104, 111, 115, 131, 136, 139, 140, 145, 182, 188, 189, 195, 196
processing, 147
processing stage, 118
transportation, 55, 115, 140
Raw Material Flow (RMF), 6, 103, 206
approach, 147
method, 56–58, 61, 93, 116, 118, 147
Renewable
electricity, 1, 23
generation, 7, 40, 42, 73, 103, 105, 136
generation environmental hazards, 3
generation plant, 139
mixes, 33, 35, 37, 39
energy, 11, 31, 35, 87, 95, 145, 165
carbon emissions, 10
deficiency, 185
effect assessments, 105, 138
elements, 205
environmental impact assessment, 45
generation, 24
generation resources, 135
generation systems, 5, 103, 111, 135, 139
generation technologies, 135
plants, 1, 5, 8, 115, 122, 140, 150
production, 9, 74
resources, 132
sources, 17, 48, 178
plants, 1, 2, 6, 9, 11, 31, 40–42, 103, 129, 156, 205
power
generation, 7, 115, 140
plant, 1, 2, 4, 7, 10, 12, 14, 17, 31, 40, 41, 104, 109, 116, 118, 121, 136, 138, 146, 147, 150, 180, 206, 207
resource, 11

systems, 11
technologies, 8, 11, 14, 17, 31
production technologies, 137
resources, 41, 46, 156
sources, 74, 94, 209
systems, 10, 135, 209
Renewable energy system (RES), 1, 2, 5, 8–11, 13, 31, 45–47, 104, 136
impacts, 8, 10, 45
LCA, 17
Renewable energy technology (RET), 5, 8–11, 13, 40, 74, 104, 136, 161
LCA analysis, 14
operation, 2, 205
Renowned ecoinvent database, 115
Research
area, 6
findings, 93
gaps, 7, 11, 45, 205, 206
groups, 104, 137, 180
outcomes, 51, 108, 138, 164
work, 1, 75, 77, 80, 102, 105, 138, 177, 207
Reservoir
hydropower plants, 84
plants, 102
Robust control operation, 4
Routing
excess electricity, 183
excess energy, 3, 6, 161, 163, 174, 179, 183
excess power, 4, 164
unit, 163

S

Sensitivity analysis, 2, 18, 23, 40, 49, 54, 58, 68, 71, 102, 129, 147, 179, 181, 183, 188, 193, 198–200, 208, 209
electricity generation, 29
hydropower plants, 24
SimaPro software, 6, 47, 56, 80, 103, 105, 132, 138, 145, 160, 207
programs, 6
version, 75, 85, 115, 182, 191
Smart energy sharing, 162
Smart houses, 164, 165, 171, 173, 184–186
energy demand, 168

PV generation units, 165
PV panels, 184
Solar
cells, 47, 104, 137
collector, 39, 47, 49, 52, 61, 65, 68, 69, 71, 206, 207
energy, 47, 52, 135
environmental impacts, 54, 138
home system, 164
impacts power plants, 17
irradiation, 172
modules, 61
panel, 23, 47, 65
plant installations, 23
plants, 18, 42, 55, 154
power, 17
generation, 109
plant, 137–139, 147–151
station, 22
systems sensitivity assessment, 23
production, 172
scale, 208
scale rates, 199
silicon, 18
systems, 37, 69, 71, 72
technologies, 3, 9, 48, 49, 54, 55, 71, 72, 207
Stable operation, 165, 176
Stratospheric ozone depletion, 90
Superior environmental performance, 47, 65
Surplus energy, 164, 185
Sustainability assessment, 37
electricity generation, 35
environmental, 24
Sustainability indicators, 7
Sustainable
energy
generation, 48
production, 104
technologies, 48
operation, 187
renewable
electricity generation, 7, 206
energy generation, 9
power plant, 35
Systematic assessment, 147

Index

Systematic LCA, 2, 71, 75, 109, 138, 206
   analysis, 132, 160, 206
   approach, 5, 48, 182
   for assessing environmental profiles, 108, 138

T

Techno-economic analysis, 102, 180, 183
Terrestrial
   acidification, 75, 90, 182, 191, 200
   ecosystems, 88, 191, 192
   ecotoxicity, 15, 42, 56, 75, 91, 136, 147, 148, 154, 194
   eutrophication, 41, 61, 63, 72
Tool for the Reduction and Assessment of Chemical and other Environmental Impacts (TRACI), 15, 103, 105, 118, 123, 131, 132, 206
   analysis, 124
   method, 40, 116, 119, 127
Transportation
   distance, 131
   type, 29

U

Uncertainty analysis, 3, 4, 58, 72, 77, 78, 87, 98, 105, 108, 109, 117, 118, 124, 127, 138, 139, 146, 152

United Nations Framework Convention on Climate Change (UNFCCC), 180
Utilizing energy, 205, 207

W

Waste
   emissions, 13, 84, 115, 140
   heat emissions, 129
   management, 49, 54, 71, 105, 138, 189, 206, 209
Water consumption, 73, 85, 89, 182, 192, 194, 195, 199, 200, 205, 206
Wind
   electricity generation systems, 39
   electricity production, 26
   energy, 41, 104, 109, 136
   energy plants, 37
   locations, 140
   penetration, 26
   plants, 27, 208
   power, 37
      environmental effects, 9
      generation, 27
      plants, 24, 26, 27, 135–140, 147–149, 151, 153, 207
      plants environmental impacts, 27
      plants impacts, 42
   turbine, 9, 27, 41, 104, 137
   turbine manufacturing, 149

Printed in the United States
by Baker & Taylor Publisher Services